MW00760196

94

Structure and Bonding

Springer-Verlag Berlin Heidelberg GmbH

Liquid Crystals I

Volume Editor: D.M.P. Mingos

With contributions by
M.A. Athanassopoulou, M.A. Bates,
S.J. Clark, J. Crain, W. Haase, G.R. Luckhurst,
B.I. Ostrovskii, M.R. Wilson

Springer

In references Structure and Bonding is abbreviated
Struct.Bond. and is cited as a journal.

Springer WWW home page: HTTP://www.springer.de

ISSN 0081-5993
ISBN 978-3-662-14716-0   ISBN 978-3-540-68305-6 (eBook)
DOI 10.1007/978-3-540-68305-6

CIP Data applied for

Typesetting: Scientific Publishing Services (P) Ltd, Madras
Production editor: Christiane Messerschmidt, Rheinau
Cover: Medio V. Leins, Berlin
SPIN: 10649377   66/3020 – 5 4 3 2 1 0 – Printed on acid-free paper

Volume Editor

Prof. D. Michael P. Mingos
Department of Chemistry
Imperial College of Science, Technology and Medicine
South Kensington
London SW7 2AY, UK

Editorial Board

Editorial Statement

It is difficult to accept that the first volume of *Structure and Bonding* was published more than 30 years ago. The growing series of volumes with their characteristic green and white covers has provided a personal landmark throughout the scientific career of many of us. When you visit a new library, the green and white bands not only guide you to the inorganic section, but also provide some stimulating reading. For some of us, our views on ligand field theory, hard and soft acids and bases, and bonding in cluster compounds were structured by articles in these volumes. The series has almost reached one hundred, and it is perhaps appropriate to reconsider its aims and scope.

The distinguished original Editorial Board consisted of C.K. Jørgensen, J.B. Neilands, R.S. Nyholm, D. Reinen and R.J.P. Williams – names that we clearly associate with the renaissance of modern inorganic chemistry. Not surprisingly the Preface in the first volume showed the foresight and imagination that one would expect of such pioneers. They argued that a new review series was justified because "A valuable service is performed by bringing together up-to-date authoritative reviews from the different fields of modern inorganic chemistry, chemical physics and biochemistry, where the general subject of chemical bonding involves (usually) a metal and a small number of associated atoms. . . . We are especially interested in the role of the 'complex metal-ligand' moiety. . . and wish to direct attention towards borderline subjects. . . . We hope that this series may help to bridge the gaps between some of these different fields and perhaps provide in the process some stimulation and scientific profit to the reader".

Since that time progress in all the scientific disciplines that they highlighted in their Preface has been enormous, and the volume of primary publications has increased exponentially. At the same time powerful new tools have become available to chemists to enable them to tackle problems that were unthinkable only a few years ago. They were right to stress the importance of inter-disciplinary studies, and indeed the growth of bioinorganic chemistry has been so dramatic but it now justifies its own review series, *Topics in Biological Inorganic Chemistry*, which has been initiated by some of the former *Structure and Bonding* editors. In the flush of success associated with the development of ligand field theory, the founding editors perhaps stressed the importance of the metal-ligand bond too much for modern tastes. Inorganic chemists are now freer to exploit all parts of the periodic table and tackle problems in materials chemistry, biology and medicine. Also the traditional barriers between organic and inorganic chemistry now seem an irrelevance. This catholic approach to the subject is held together by the basic belief that an understanding of the structure and dynamics of the chemical interactions at the molecular level will lead to great insights into the observations made in the laboratory or observed in nature.

We expect the scope of the *Structure and Bonding* series to span the entire periodic table and address structure and bonding issues wherever they may be relevant. Therefore, it is anticipated that there will be reviews dealing not only with the traditional areas of chemical bonding based on valence problems and dynamics, but also nanostructures, molecular electronics, supramolecular structure, surfaces and clusters. These represent new and important developing areas of chemistry, but others will no doubt also emerge. Physical and spectroscopic techniques used to determine, examine and model structures will fall within the purview of *Structure and Bonding* to the extent that the focus is on the scientific results obtained and not on specialist information concerning the techniques themselves. Issues associated with the development of bonding models and generalisations that illuminate not only the structures of molecules, but also their reactivities and the rates of their chemical reactions will also be considered relevant.

Previously, *Structure and Bonding* brought together not only volumes dealing with specific topics, but also published unsolicited articles. In the future, we shall only be publishing volumes on specific topics. It is our goal to provide the scientific community with thematic volumes that contain critical and timely reviews across a wide range of subjects. However, all members of the Editorial Board welcome suggestions for future volumes from readers.

We join Springer-Verlag in thanking the editors now leaving the Editorial Board, M.J. Clarke, J.B. Goodenough, C.K. Jørgensen, G.A. Palmer, P.J. Sadler, R. Weiss, and R.J.P. Williams, whose efforts over the last decades served to maintain the high standards of the series.

Therefore, with the twenty-first century near at hand, we hope that you will agree that the change we are making not only represent a cosmetic change of cover, but a change of emphasis which accurately reflects the way in which the science is developing.

Allen J. Bard, Austin
Ian G. Dance, Sydney
Peter Day, London
James A. Ibers, Evanston
Toyohi Kunitake, Fukuoka
Thomas J. Meyer, Chapel Hill
D.M.P. Mingos, London
Herbert W. Roesky, Göttingen
Jean-Pierre Sauvage, Strasbourg
Arndt Simon, Stuttgart
Fred Wudi, Los Angeles   October 1998

Preface

The liquid crystalline state may be identified as a distinct and unique state of matter which is characterised by properties which resemble those of both solids and liquids. It was first recognised in the middle of the last century through the study of nerve myelin and derivatives of cholesterol. The research in the area really gathered momentum, however, when as a result of the pioneering work of Gray in the early 1970's organic compounds exhibiting liquid crystalline properties were shown to be suitable to form the basis of display devices in the electronic products.

The study of liquid crystals is truly multidisciplinary and has attached the attention of physicists, biologists, chemists, mathematicians and electronics engineers. It is therefore impossible to cover all these aspects fully in two small volumes and therefore it was decided in view of the overall title of the series to concentrate on the structural and bonding aspects of the subject. The Chapters presented in these two volumes have been organised to cover the following fundamental aspects of the subject. The calculation of the structures of liquid crystals, an account of their dynamical properties and a discussion of computer simulations of liquid crystalline phases formed by Gay-Berne mesogens. The relationships between molecular conformation and packing are analysed in some detail. The crystal structures of liquid crystal mesogens and the importance of their X-ray scattering properties for characterisational purposes are discussed. These are followed by chapters on columnar phases, dimeric liquid crystals, twist grain boundary phases and metallomesogens. I should like to thank the authors for providing such a detailed analysis of these important aspects of the subject and doing it in a way that I think will be intelligible to the general reader. I should also like to thank Dr. John Seddon for assisting me in making the choice of authors and topics for this volume.

For the general reader the following brief summary and basic definitions may provide a helpful introduction:

A compound which displays liquid crystal properties is referred to as a *mesogen* and said to exhibit *mesomorphism*. Liquid crystals may be considered either as disordered solids or ordered liquids, and their properties are very dependent on temperature and the presence or absence of solvent. In *thermotropic liquid crystals* the phases of the liquid crystals may be observed to change as the temperature is increased. In *lyotropic liquid crystals* the ordered crystalline state is disrupted by the addition of a solvent, which is very commonly water. For these systems temperature changes may also be responsible for phase changes, but only when used in combination with solvent.

The conventional liquid state is described as the *isotropic phase*. The temperature at which the compound passes from the solid phase into a mesophase is described as the *melting point* and the transition temperature between a mesophase and an isotropic liquid is described as the *clearing point*.

The mesophases of thermotropic liquid crystals are described as *calamitic* if the constituent molecules are rod-like and *columnar*, if the constituent molecules, which often have a disc like shape(discotic), stack into columns.

The simplest mesophase is the nematic phase. It is very fluid and involves highly disordered molecules having only short-range positional order, but with the molecules preferentially aligned on average in a particular direction (the director). If the constituent compound is racemic then it is possible to form a phase from the enantiomerically pure compound which is a *chiral nematic phase.*

Smectic phases are more highly ordered than nematic phases, and with an ordering of the molecules into layers. There are a number of different smectic phases which reflect differing degree of ordering. *Crystal smectic phases* are characterised by the appearance of inter-layer structural correlations and may in some cases be accompanied by a loss of molecular rotational freedom.

Columnar mesophases are observed generally for disc shaped molecules and in these phases the molecules show some orientational order with respect to the short molecular axis. Nematic phases which are directly related to calamitic nematic phases are observed and are similarly very fluid. Many of the columnar phases have symmetry resulting from ordered side-to-side packing— with or without ordering in the columns. In the disordered hexagonal phase the molecules define columns which are packed together in a hexagonal manner. Within the columns there is a variable degree of ordering. The ordered phases resemble the crystal smectic phases. Rectangular, oblique and tilted phases have also been identified.

Micelles are formed when surfactant molecules are dissolved in water. The precise concentration for micelle formation is described as the *critical micelle concentration*. Micelle formation results because of the disparate nature of the two ends of the surfactant molecules, i.e. there are hydrophobic and hydrophilic ends. If the concentration of surfactant is increased above the critical concentration then the density of micelles becomes sufficiently large that they begin to form ordered arrays. These ordered arrays correspond to *lyotropic liquid crystals*. Different mesophases result from different micelle types. Spherical micelles in general give rise to cubic close packed phases. Rod-like micelles give rise to hexagonal mesophases and disc mesophases give rise to lamellar phases. A variety of further phases are based on bilayer structures, ranging from the flat lamellar phases, to the highly convoluted bicontinuous phases, many of cubic symmetry, in which the interface is curved.

Many polymers are capable of forming mesophases in either aqueous or non-aqueous solvents. Furthermore, liquid crystal phases may form for pure block-copolymers through a tendency for the different polymer blocks to separate on a microscopic scale.

Hopefully with this brief introduction the reader will be able to appreciate fully the chapters which follow and which have been written by experts in the field. The complexity and beauty of the liquid crystalline phase has attracted many able scientists and the applications of liquid crystals in the electronics industry have provided a secure funding base for the subject. This is therefore still a field which is expanding rapidly and many research avenues remain to be explored by newcomers. Perhaps after reading these volumes of *Structure and Bonding* you will be tempted to join this exciting endeavour.

D.M.P. Mingos

Contents

Contents of Volume 95

Liquid Crystals II

Volume Editor: D. M. P. Mingos

Calculation of Structure and Dynamical Properties of Liquid Crystal Molecules

Jason Crain[1], S. J. Clark[2]

[1] Department of Physics and Astronomy, The University of Edinburgh,
   Kings Buildings, Mayfield Road, EH9 3JZ Scotland, UK, *Email:* j.crain@ed.oc.uk
[2] Department of Physics, Durham University, Science Labs, South Road, Durham
   DH1 3LE, UK, *Email:* s.j.clark@durham.ac.uk

In this article the motivation for studying the structure and dynamics of liquid crystal molecules using quantum mechanical electronic structure techniques will be outlined. Recent theoretical advances as well as algorithmic developments in ab initio calculations on large molecular systems will be introduced and results which elucidate the nature of structure and bonding in liquid crystals will be reviewed.

Keywords: Ab initio, Structure-function relationships, Molecular flexibility, Density functional theory, Molecular vibrations

Structure and Bonding, Vol. 94
© Springer Verlag Berlin Heidelberg 1999

1
Introduction

1.1
The Structure-Property Problem

Structure and bonding in liquid crystals is a very rich and diverse field which exists at the interface between modern physics and chemistry. At one extreme, it refers to large scale, often mesoscopic structures (such as cholesteric and Twist-Grain-Boundary phases [1]) and to long-range interactions. This is the intermolecular regime. In the opposite limit, the intramolecular regime, the important properties are geometry, flexibility and the detailed electronic structure of individual molecules. Ultimately, these molecular properties determine the condensed phase behaviour. In some cases, however, the inter- and intramolecular regimes are separable in that it is possible to discuss the origin of several liquid crystalline phases using very idealised representations of molecular structure and bonding. In this way it is possible to explore some very fundamental properties of the liquid crystal state which should be common to all mesogenic materials [2, 3]. These simplified molecular models are also very efficient to implement on a computer and can be used to study a range of material properties using standard computational methods [4]. Some of the simplest and most common molecular models of liquid crystals are those which represent the molecules as rigid rods, ellipsoids or spherocylinders (cylinders truncated at each end by hemispheres). There is clearly no internal molecular structure present in these models, although the requirement that the molecules are hard objects implies that no two molecules can interpenetrate. The molecular origin of this excluded volume interaction lies in the fact that electron clouds on adjacent molecules cannot overlap. Therefore repulsive forces dominate at short separations for this reason. This short-ranged repulsion is, in itself, sufficient to give rise to a variety of orientationally ordered phases in these systems with nematic and smectic phases being observed as has been demonstrated in a variety of simulations [2] motivated by the early theoretical work of Onsager [5]. Since there is no source of attractive interactions in these models, phase behaviour is determined only by molecular shape (which is usually fixed) and density. Temperature does not enter into it and these systems do not exhibit thermotropic transitions. Other idealised molecular models have been developed and refined in order to incorporate anisotropic attractive interactions between liquid crystal molecules. One of the most well-known is the Gay-Berne (GB) model where the interaction is based on a generalisation of the isotropic Lennerd-Jones potential [6]. The parameters in this model refer to gross features of intramolecular shape and anisotropy of attractive interactions. As with the rigid rod case, no intramolecular dynamical degrees of freedom are present. A wide variety of liquid crystalline phases have also been observed using the GB description [7]; however it is difficult to relate the parameters which enter the model to details of the molecular structure and bonding.

Alternatively, it is possible to study in detail the structure and properties of the individual complex molecules which form liquid crystals and to try to relate these to observed condensed-phase material properties. The need to develop new molecular materials for applications as well as advances in synthesis and characterisation are all providing a major new motivation to develop a deeper understanding of the relationships between the inter- and intramolecular regimes. The importance of relating the properties of materials to those of their individual molecular constituents is a well-known and long standing challenge. However, in practice, this structure-property problem is exacerbated in the case of liquid crystals for several reasons. First, despite the fact that liquid crystals are exotic phases of matter, liquid crystallinity is a relatively common phenomenon and over ten thousand materials are known to exhibit one or more liquid crystalline phases. Also new mesogens are regularly being synthesised [8, 9]. These compounds are drawn from a wide variety of chemical classes which means that the scope of the structure-property problem for liquid crystals is very wide indeed. A second practical problem arises in the context of applications where "optimal" molecular properties depend on the details of device architecture. This will be considered in more detail in a later section. Third, most molecules which form liquid crystals are relatively large, containing at least several tens of atoms in total, several different atomic species and a huge number of valence electrons. The molecules therefore tend to be flexible (at least in part) which gives rise to considerable conformational variety. The difficulty is that many of the most important intramolecular properties such as dipole moment, quadrupole moment and polarisability (derived directly from intramolecular electronic bonding) are directly coupled to molecular shape and therefore may change considerably with conformation. There exists, therefore, another smaller-scale, structure-property problem even for a single isolated liquid crystal molecule. A schematic illustration of interactions in mesogens and their molecular electronic origin is shown in Fig. 1.

1.2
Objectives for Theory and Simulation

In principle some of these structure-property problems are surmountable by reliable experiments. Insight into molecular structure can be obtained though diffraction [10] and Nuclear Magnetic Resonance (NMR) measurements [11]. Vibrational spectroscopy can give some information on bond strength provided mode frequencies can be unambiguously assigned [12]. Molecular flexibility is much harder to determine quantitatively though NMR [11] can provide some indirect data on the probability of conformers in different phases. Molecular dipoles are accessible as conformational averages by a variety of techniques but tend to be measured in solution as opposed to in gas phase and the influence of the solvent may complicate interpretation. Of course it is often desirable when designing new materials to have a sense of how a material will behave before it is synthesised. In this regard, experiment does not really afford a predictive capability and reliable computer simulation can, in principle, take on increased significance.

Computer simulations therefore have several inter-related objectives. In the long term one would hope that molecular level simulations of structure and bonding in liquid crystal systems would become sufficiently predictive so as to remove the need for costly and time-consuming synthesis of many compounds in order to optimise certain properties. In this way, predictive simulations would become a routine tool in the design of new materials. Predictive, in this sense, refers to calculations without reference to experimental results. Such calculations are said to be from *first principles* or *ab initio*. As a step toward this goal, simulations of properties at the molecular level can be used to parametrise interaction potentials for use in the study of phase behaviour and condensed phase properties such as elastic constants, viscosities, molecular diffusion and reorientational motion with maximum specificity to real systems. Another role of ab initio computer simulation lies in its interaction

Inter/Intramolecular Interaction	Molecular Electronic Origin	Molecular Models
Short-ranged repulsive	No overlap of molecular electronic charge distributions (Pauli principle)	Rigid objects (rods, ellipsoids etc) 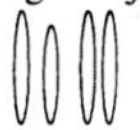
Short-ranged attractive	Molecular polarisability anisotropy and high order multipoles	Gay-Berne
Long-ranged attractive	Dipolar forces	Rigid rod or GB + dipolar term  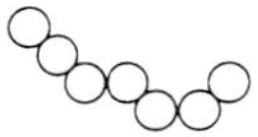
Coupling between structure and orientational order	Energy cost of distorting electron distribution	quantum mechanical atomistic models
Molecule/surface interactions (molecular basis for anchoring)	electrostatic, chrage transfer,	empirical potentials fully ab initio not yet performed

Fig. 1. Summary of interactions in LCs and their molecular electronic origin

with experiment. Predictive, parameter free methods can be very useful tools in interpreting experimental data from a variety of sources such as NMR, X-ray, Raman, ESR and others. Furthermore, simulations can be used to investigate properties which are difficult to explore experimentally or to separate complex effects into individual contributions.

The maintenance of a connection to experiment is essential in that reliability is only measurable against experimental results. However, in practice, the computational cost of the most reliable conventional quantum chemical methods has tended to preclude their application to the large, low-symmetry molecules which form liquid crystals. There have however, been several recent steps forward in this area and here we will review some of these newest developments in predictive computer simulation of intramolecular properties of liquid crystals. In the next section we begin with a brief overview of important molecular properties which are the focus of much current computational effort and highlight some specific examples of cases where the molecular electronic origin of macroscopic properties is well established.

2
Molecular Properties

2.1
Liquid Crystal Molecules

2.1.1
Generic Molecular Structures

There are two main types of thermotropic liquid crystals which are distinguished on the basis of molecular shape. The distinction is between disc-shaped (discotic) and rod-shaped (calamitic) molecules. Discotics are flat disc-like molecules consisting of a core of adjacent aromatic rings. This allows for mesophases based on columnar ordering. Rod-shaped molecules have an elongated, anisotropic geometry which allows for preferential alignment along one spatial direction. We will focus on rod-like (calamitic) liquid crystals as these have been studied in great detail for many years. Most calamitic liquid crystal molecules are comprised of several well-defined building blocks. These are the core, usually comprising two or more ring systems joined by a linking group and possibly incorporating lateral substituents on the rings. An illustration of this type of molecular structure is shown in Fig. 2. At either end are terminal groups of which one is usually a long alkyl, acyl or alkoxy chain. As these are the constituents of mesogens it is imperative to describe accurately their electronic structure and to understand the influence of substituents. Examples of liquid crystals having this general structure type are the cyano-biphenyl homologous series which exhibit nematic and smectic mesophases depending on the length n of the alkyl chain. Correlations between phase transition temperatures between isotropic and nematic phases (T_{NI}) and alkyl chain length have been well-documented for several years [13]. Normally T_{NI} exhibits a decreasingly pronounced alternation, depending on whether n is

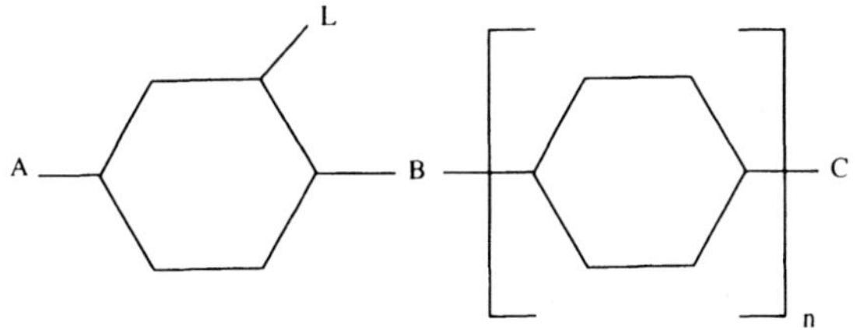

Fig. 2. The generic molecular structure of calamitic liquid crystals illustrating the semi-rigid core fragments, the positions of the end-groups (C and A), linking groups (B) and, possibly, laterally substituted groups (L)

odd or even, for increasing n. Other properties exhibiting odd-even effects include birefringence [14], helical pitch [15], alignment [16] and dielectric properties [17]. Very fundamental studies of the nematic to smectic A tricritical point have exploited these well-known trends in structure adoption as a function of chain length [18]. The formation of smectic phases has also been explained purely in terms of molecular shape arguments [19]. The melting points of liquid crystals tend to exhibit least correlation with molecular structure. This is because the stability of a solid phase is governed by short-ranged attractive forces which are not related in a simple way to molecular structure.

Other more exotic types of calamitic liquid crystal molecules include those having chiral components. This molecular modification leads to the formation of chiral nematic phases in which the director adopts a natural helical twist which may range from sub-micron to macroscopic length scales. Chirality coupled with smectic ordering may also lead to the formation of ferroelectric phases [20].

2.1.2
Electronic Structure and Orientational Order

In order to have a well-defined shape (such as rod-like or plank-like) as assumed in the preceding section, it is implicit that molecules must exhibit a certain degree of rigidity. For example, long-chain fatty acids (*n*-alkanoic acids) can exist in very extended conformers but are very flexible and do not form thermotropic liquid crystalline phases. The rigidity which defines the molecular shape ultimately has its origin in the molecular electronic structure. For example, the introduction of double bonds restricts internal rotational motion about the sp^2 carbon bonds and the resulting compounds have a sufficiently rigid core structure to form a nematic phase [13]. Linking groups (shown as component B in Fig. 2) which preserve linearity and which are also unsaturated extend conjugation between the rings and enhance the rigidity.

Geometrical shape is by no means the only molecular property which governs phase stability and behaviour and there are numerous examples of cases in which molecules of similar shape form condensed phases with very different properties. These differences are attributed to subtle chemical changes which alter the electronic structure. For instance, the transition temperatures and phase stability of fluorinated terphenyls having identical chemical compositions can be dramatically altered simply by changing the location of the fluorine substituent [21]. This is illustrated for two cases in Fig. 3.

The compound in Fig. 3b exhibits two smectic phases (Sm_C and Sm_A) in addition to nematic, whereas the compound in Fig. 3a exhibits only a nematic phase. The substitution of an alkoxy for an alkyl tail is known to shift phase transition temperatures considerably. In the cyano-biphenyls (Fig. 4), substitution of an alkoxy tail raises the melting point from 24 to 48 °C and T_{NI} from 35 to 68 °C [22].

Small chemical changes to the tolane family of mesogenic molecules are also known to bring about major changes in phase behaviour [22]. Two examples are shown in Fig. 5 where subtle changes in the tail can eliminate the nematic phase.

In some cases, chemical substituents can bring about unusual *monotropic* liquid crystalline phases which only exist upon heating or cooling. The diphenyl-diacetylenes are examples in this category [22]. Early theoretical connections between molecular electronic structure and orientational order

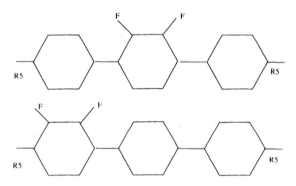

Fig. 3a,b. Two fluorinated terphenyl molecules: **a** system exhibiting only a nematic phase with T_{NI} = 120 °C, freezing at T = 60 °C; **b** system exhibiting the two intervening smectic phases following the phase sequence K $\xrightarrow{81}$ S_c $\xrightarrow{115.5}$ S_A $\xrightarrow{131.5}$ N $\xrightarrow{142}$. R5 refers to terminal pentyl tails

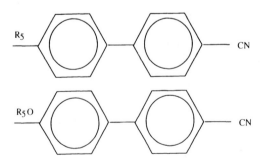

Fig. 4a,b. The effect of substitution of an alkoxy tail for an alkyl tail in cyanobiphenyls: **a** transition sequence is K $\xrightarrow{24}$ N $\xrightarrow{35.3}$ *I*; **b** transition sequence is K $\xrightarrow{48}$ N $\xrightarrow{68}$ *I*

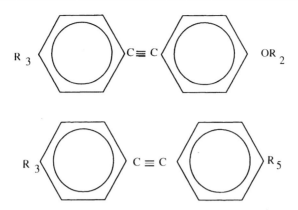

Fig. 5a,b. Tolane compounda: **a** melts at 39 °C and is not mesogenic;. **b** differs only in one tail substituent melts to a nematic phase at 61 °C which persists until 89 °C

formed part of the foundations of the Maier-Saupe mean field theory of the nematic phase [23–25]. According to the original formulation the dipole-dipole part of anisotropic dispersion forces gives rise to an average orientational energy u such that

$$u \propto \tfrac{1}{2}(3\cos^2 \theta_1 - 1)S \tag{1}$$

where S is the order parameter and θ_i is the angle made by the long molecular axis relative to the director. The polarisability anisotropy $\Delta\alpha$ determines the strength of these dispersion forces. Molecular theories of lower symmetry smectic-C phases have also been formulated involving quadrupolar or dipole-induced dipole interactions [26]. Some of these models attribute the tendency to form tilted phases to the presence of large lateral molecular dipoles [27] or to quadrupolar perturbations of the smectic-A structure [28].

A simple example of how molecular electronic structure can influence condensed phase liquid crystalline properties exists for molecules containing strongly dipolar units. These tend to exhibit dipolar associations in condensed phases which influence many thermodynamic properties [29]. Local structural correlations are usually measured using the *Kirkwood factor g* defined as

$$g = \sum_{n \neq m} \langle P_1(\cos(\phi_{nm})) \rangle \tag{2}$$

In this expression ϕ_{nm} is the angle between molecular dipoles on molecules n and m and P_1 is the first Legendre polynomial. A value of $g < 1$ corresponds to a preference for anti-parallel dipolar order while $g > 1$ arises from local parallel order. The type and degree of ordering is sensitive to molecular structure though it is generally found that terminal polar substituents tend to align in anti-parallel whereas laterally substituted materials exhibit parallel correlations [30]. The strength and location of the molecular dipole moments have also been found to play a role in determining phase stability in liquid crystals [31, 32].

Another example of the coupling between microscopic and macroscopic properties is the flexo-electric effect in liquid crystals [33] which was first predicted theoretically by Meyer [34] and later observed in MBBA [35]. Here orientational deformations of the director give rise to spontaneous polarisation. In nematic materials, the induced polarisation is given by

$$\hat{P} = e_{11}\hat{n}(\nabla \cdot \hat{n}) + e_{33}(\nabla \times \hat{n}) \times \hat{n} \tag{3}$$

where $\hat{n}$ is the director and e_{11} and e_{33} are called flexoelectric moduli. They have units of charge/length and are defined by

$$e_{11} = 2\left(\frac{a}{b}\right)^{1/3} \mu_{\parallel} K_{11} O_0 N^{1/3} / k_b T \tag{4}$$

$$e_{33} = 2\left(\frac{a}{b}\right)^{1/3} \mu_{\perp} K_{33} O_0 N^{1/3} / 2 k_B T \tag{5}$$

where $\mu_{\parallel}$ and $\mu_{\perp}$ are the longitudinal and transverse components of the molecular dipole and N is the number density of molecules. The parameters a, b and O_0 refer to molecular shape which is usually bent or boomerang shaped as shown in Fig. 6.

The factors K_{ii} are elastic constants for the nematic phase and k_B is the Boltzmann constant. Therefore a combination of molecular electronic structure, orientational order and continuum elasticity are all involved in determining the flexoelectric polarisation. Polarisation can also be produced in the presence of an average gradient in the density of quadrupoles. This is

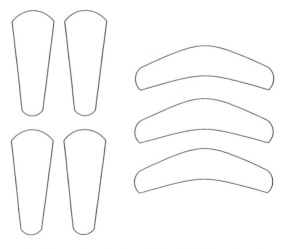

Fig. 6.   The two generic shapes of molecules which exhibit flexoelectric polarisation under distortion of the equilibrium director distribution

independent of molecular shape though there is an implicit dependence of the single molecule quadrupole on molecular shape.

A variety of theories have been developed to link molecular structure and bonding to other condensed phase continuum properties such as elastic constants [36] including biaxial nematic phases [37], viscosities [38] and nematic flow alignment under shear [39]. In many of these cases the molecular properties which enter the formulations are the aspect ratio, orientational order parameters, inertia tensors and intermolecular interaction energy. Recent experiments involving the freezing of successive liquid crystal layers have revealed evidence for exotic attractive forces [40] (short-ranged exponential and thermal Casimir) which help to bind liquid crystals. The molecular electronic origin of these remains to be explored.

2.2
Structure, Bonding and Device Applications

2.2.1
Overview

Much current research on molecular design of liquid crystalline materials is driven by the requirement of specific device applications. Liquid crystalline materials must show a wide mesogenic range, switch quickly between on and off states in applied fields using relatively low voltages, and their orientation on solid substrates must be controllable and as defect-free as possible. As discussed in the previous section, the stability and material properties of liquid crystals have their origins in the structure and bonding of individual molecules. In some cases these relationships are reasonably straightforward but more often they are relatively complex. Below we explore these relationships for some of the most important and exploitable material properties of liquid crystals.

2.2.2
Material Stability

Commercially viable devices must be capable of functioning over a relatively wide range of external conditions. For liquid crystal devices a stringent requirement is that they can be operated over a reasonably wide temperature range and exhibit a high degree of chemical and photochemical stability. Clearly this means that the range of the liquid crystalline phase which is exploited in the device be relatively large and close to room temperature.

Currently, theories are not yet able to predict the transition temperatures based on molecular structure of the constituent molecules. However, for several compounds there is considerable empirical data relating the transition temperature between isotropic and nematic phases (T_{NI}) to molecular structure. Higher T_{NI} implies greater nematic stability. For example, it is now well established that in Shiff's base molecules [13] the incorporation of increasingly polarisable substituents increases T_{NI} in accord with the theory of Maier and Saupe [23]. The influence of thermal stability of the nematic phase has now been studied for a range of different substituents and reported in [30, 31]. In general, considerable sensitivity to substituents is found and qualitative explanations are often formulated in terms of the relative contributions of the substituents to electronic polarisability and steric effects.

2.2.3
Optical Properties

It is the optical properties of liquid crystals which have brought and kept them at the forefront of molecular electronics. The most important of these properties include the birefringence and refractive index. The birefringence of a material is defined as the difference in the indices of refraction for optical radiation having orthogonal polarisations. For a typical nematic liquid crystal this difference may be between 0.1 and 0.2. The details of the molecular electronic structure and energy levels determine the optical properties. In particular, the electronic structure of the ring units (Fig. 2) in liquid crystals dominates the important macroscopic properties such as polarisability, optical absorption and refractive indices. The rings usually comprise either σ-bonded cyclohexane units or phenyl units containing a mixture of σ and π-bonded carbon atoms. Molecules containing both types of ring system are also possible. In saturated σ-bonded systems optical absorption is governed by $\sigma \rightarrow \sigma^*$ transitions which occur in the ultraviolet part of the spectrum, meaning that these materials will be transparent in the visible. The $\sigma \rightarrow \sigma^*$ transitions tend to have high oscillator strength but absorption is isotropic. These transitions therefore contribute mostly to the refractive indices of liquid crystals and rather less to their birefringence. By contrast, the $\pi \rightarrow \pi^*$ transition process in phenyl-containing compounds gives absorption at much longer wavelengths. This absorption is highly anisotropic and these transitions dominate the birefringence while having a lesser effect on the refractive index. The electronic transitions in benzene systems have been the subject of extensive theoretical studies using molecular orbital theories [41]. It is expected that the absorption spectrum of a liquid crystal molecule containing a single phenyl ring would exhibit similarities to that of an isolated benzene molecule depending on the degree of conjugation. Various theoretical and semi-empirical models have been developed to describe electronic anisotropic optical properties and absorption mechanisms in liquid crystals, some of which take explicit account of the $\sigma \rightarrow \sigma^*$ and $\pi \rightarrow \pi^*$ transitions [42]. Polarised absorption spectra for a variety of liquid crystals can be found in [22] for biphenyl-containing mesogens though other ring systems can be synthesised [13, 30].

In addition to the sensitivity to ring structure, slight alterations of the hydrocarbon chain may lead to dramatic differences in electro-optic performance in chiral compounds [43]. For example, in some electroclinic Sm-A materials it has been reported that if the chain is shortened or if a double bond is localised at the end of the hydrocarbon chain, the tilt angle, electroclinic coefficient, and switching time are significantly suppressed [43].

As a result of the unique molecular electronic properties and geometrical structure, liquid crystals also tend to be optically nonlinear materials meaning that their properties are influenced by applied optical fields. These nonlinear optical properties are derived from higher-order hyperpolarisabilites and they have begun to be exploited in a variety of applications in switching and mixing. These have been reviewed in detail in [22].

2.2.4
Orientational Response to Applied Fields

The response of liquid crystal molecular orientation to an electric field is another major characteristic utilised for many years in industrial applications [44] and more recently in studies of electrically-induced phase transitions [45]. The ability of the director to align along an external field again results from the electronic structure of the individual molecules.

In most devices the liquid crystal molecules are confined between two thin walls which act as capacitor plates. This allows the determination of the dielectric properties of the liquid crystalline material through the simple relations

$$\varepsilon_{\|,\perp} = \frac{C_{\|,\perp}}{C_c} \tag{6}$$

where C_c is the empty-cell capacitance and $\varepsilon_{\|,\perp}$ is the dielectric constant parallel and perpendicular to the director.

Usually it is the so-called dielectric anisotropy, defined as

$$\Delta\varepsilon = \varepsilon_\| - \varepsilon_\perp \tag{7}$$

which is of most significance in liquid crystal-based applications. For display devices this is a very important material property as it determines the degree of coupling between applied electric fields and the liquid crystalline material. The theory of Maier and Maier [46] illustrates how the macroscopic dielectric anisotropy is connected to the details of the molecular electronic structure. Assuming that the liquid crystalline molecules are centres of anisotropic polarisability the dielectric anisotropy is connected to the molecular dipole moment and the polarisability through

$$\Delta\varepsilon \propto \left[\Delta\alpha - C\mu^2\left(1 - 3\cos^2\beta\right)\right]S \tag{8}$$

where S is the order parameter characterising the liquid crystalline phase. The quantity β represents the angle between the dipole moment μ and the long molecular axis and $\Delta\alpha$ is the polarisability anisotropy of the molecule. These latter quantities are determined entirely by the electronic structure of the individual molecules and the response to macroscopic applied fields is therefore governed by a combination of intramolecular electronic structure (via $\bar{\mu}$, $\Delta\alpha$ and β and the degree of long-range order in condensed phases (via S).

It is evident that the sign of the dielectric anisotropy is not fixed by the form of this expression and that it depends on the magnitude of $\Delta\alpha$ as well as on the magnitude and orientation of the molecular dipole. In general the liquid crystal director will align such that the largest component of the dielectric constant lies along the applied field direction. This molecular electronic property therefore dictates the type of device architecture which is suitable for a given compound of a given dipole moment and polarisability. For example, in a homeotropically aligned device the liquid crystal molecules must exhibit negative dielectric anisotropy [47]. For active matrix applications high resistivity is also required in order to support high voltage without flicker. By contrast, super-twisted nematic devices normally require materials having positive dielectric anisotropy [47].

From a practical perspective a high dielectric anisotropy leads to a low operating voltage V_c according to

$$V_c = \pi\left(\frac{k'}{\Delta\varepsilon\varepsilon_0}\right) \tag{9}$$

In this expression k' represents an average elastic constant.

In the operation of ferroelectric liquid crystal devices, the applied electric field couples directly to the spontaneous polarisation $\vec{P}_s$ and response times depend on the magnitude $\vec{E} \cdot \vec{P}_s$. Depending on the electronic structure (magnitude and direction of the dipole moment as well as position and polarity of the chiral species) and ordering of the molecules $\vec{P}_s$ can vary over several orders of magnitude (3 to 1.2×10^3), giving response times in the range 1–100 μs.

2.2.5
Alignment on Substrates

An essential requirement for device applications is that the orientation of the molecules at the cell boundaries be controllable. At present there are many techniques used to control liquid crystal alignment which involve either chemical or mechanical means. However the relative importance of these two is uncertain and the molecular origin of liquid crystal anchoring remains unclear. Phenomenological models invoke a surface anchoring energy which depends on the so-called "surface director", $\hat{n}_s$. In the case where there exists cylindrical symmetry about a preferred direction, $\hat{n}_p$ the potential is usually expressed in the form of Rapini and Popoular [48]

$$F_{\text{anch}} = \tfrac{1}{2}c_0(\theta - \theta_p) \tag{10}$$

where θ and θ_p are the polar angles of $\hat{n}_s$ and $\hat{n}_p$, respectively. The quantity c_0 is a measure of anchoring strengths and it is ultimately related to bulk nematic properties and the detailed molecular electronic interactions which take place between the surface and the liquid crystal molecules.

At present there have been no attempts to explore the molecular electronic basis for liquid crystal anchoring or to calculate Rapini-Popoular coefficients for real systems using first principles methods. However, there have been a number of theoretical treatments which have suggested that the ordering of molecular dipoles in a surface layer is determined by strong dipole-quadrupole interactions, modified by the presence of the surface. Surface polarisation gives a significant contribution to the anchoring energy of nematic liquid crystals composed of strongly polar molecules some of which have yielded estimates for the anchoring strengths [49].

2.2.6
Remarks

It is clear from the forgoing discussions that the important material properties of liquid crystals are closely related to the details of the structure and bonding of the individual molecules. However, emphasis in computer simulations has focused on refining and implementing intermolecular interactions for condensed phase simulations. It is clear that further work aimed at better understanding of molecular electronic structure of liquid crystal molecules will be a major step forward in the design and application of new materials. In the following section we outline a number of techniques for predictive calculation of molecular properties.

3
Predictive Calculations of Molecular Properties

3.1
Overview

In this section we aim to introduce some of the main theoretical ideas which underlie the strategies for modelling liquid crystal molecules. It is clear that there are a very wide range of methods available and we will not attempt to be comprehensive. Instead, we will begin with a brief overview of traditional semi-empirical approaches and then progress to concentrate on treating fully predictive parameter-free calculations of molecular electronic structure and properties in some depth.

3.2
Semi-Empirical Approaches

Currently, a wide variety of methods exists for calculating the molecular structure of large liquid crystal molecules which make use of pre-determined functional forms for the interactions in a molecule and semi-empirical information to parametrise the potentials. In general the interaction terms represent the energy cost of distorting bonds and bond angles from equilibrium. These can be expressed as

$$V(\vec{r}_1, \vec{r}_2 \ldots \vec{r}_N) = \Sigma\, C_B(b - b_0)^2 + \Sigma\, C_\alpha(\theta - \theta_0)^2 + \Sigma \frac{1}{2} V_n[1 + \cos(n\phi - \delta)] \tag{11}$$

$$+ \text{ nonbonded terms}$$

where the first term is the energy required to distort a bond of length b_0 to b and C_B is a measure of the bond strength (analogous to a *spring constant*). The second term is similar to the first but describes the energy required to distort two adjacent bonds lying at an angle θ_0 to an angle θ. The third term is the energy due to a torsion ϕ in the molecule relative to some angle δ. Together, the first three terms are the total energy arising from covalent interactions. The nonbonded contributions involve coulombic and dispersion forces. In particular, if any part of the molecule is charged then they will interact via a simple Coulomb potential. Many liquid crystal forming molecules have a permanent dipole moment and therefore a dipole-dipole term is also necessary. Higher order multipole terms are thought to influence the bulk structure, but are often excluded in the above energy expression due to the lack of experimental measurements of quadrupole moments. The methods described below overcome these problems. In the above expression there is no explicit reference to the details of the molecular electronic structure. There are, instead, several free parameters which determine the importance of each contribution, these being C_B, C_α and V_n which are the empirical force constants for bond stretching, bending and torsional deformation. These methods have the advantage that they can be implemented efficiently and are therefore applicable in principle to large systems. However, the accuracy of the results of such simulations depends sensitively on the appropriateness of the parametrisation and the degree to which the parameters are transferable.

3.3
First Principles Methods and the Density Functional Principle

It is also possible to obtain a very general expression for the total energy of a molecule without any assumptions about the form of the potentials for bond deformation and without any recourse to experimental input for determining numerical values of such potentials for real systems. This amounts to obtaining an expression for the energy of the molecule in a form analogous to the Schroedinger equation for the molecular electronic wavefunction. Such a formulation, in principle, incorporates all electronic and structural degrees of freedom. There are several approaches to this problem and discussion of methodologies for implementing standard quantum chemical molecular orbital techniques can be found in [50].

In this section we introduce an alternative method which has been recently shown to provide accurate results for large liquid crystal molecules. It is based on Density Functional Theory (DFT) which was originally developed by Hohenberg and Kohn [53] and Kohn and Sham [51, 52] to provide an efficient method of handling the many-electron system. The theory allows us to reduce the problem of an interacting many-electron system to an effective single-electron problem. Hohenberg and Kohn [53] established that the electron density is the central variable and proved two theorems. First, that the total energy E of any many-electron system is a unique functional of the electron density $n(\mathbf{r})$:

$$E[n(\mathbf{r})] = F[n(\mathbf{r})] + \int V_{ion}(\mathbf{r})n(\mathbf{r})d\mathbf{r} \qquad (12)$$

where $F[n]$ itself is a functional of the density and is independent of the external potential which is usually the Coulomb potential V_{ion} due to ions (or nuclei) plus possibly other external fields. Second, for any many-electron system, the minimum value of the total energy functional is the ground state energy of the system at the ground state density.

3.3.1
Kohn-Sham Energy Functional and Equations

A key to the application of DFT in handling the interacting electron gas was given by Kohn and Sham [51] who used the variational principle implied by the minimal properties of the energy functional to derive effective single-particle Schrodinger equations. The functional $F[n]$ can be split into four parts:

$$F[n(r)] = T[n(r)] + E_H[n(r)] + E_{xc}[n(r)] + E_{ion}(R) \qquad (13)$$

where the first, second and third terms represent the kinetic energy, the Hartree energy and the exchange-correlation energy respectively. The fourth term is the interaction between positively charged ions:

$$E_{ion} = \frac{1}{2}\sum_{ij} \frac{Z_i Z_j e^2}{|R_1 - R_2|} \qquad (14)$$

where Z_i and Z_j are the valences of ions i and j. For a fixed ionic configuration R this term is a constant.

In contrast to the Hartree term which is given by

$$E_H[n(r)] = \frac{1}{2}\int\int \frac{n(r)n(r)}{|r - r'|} \, dr \, dr' \qquad (15)$$

an explicit form of other functionals, $T[n]$ and $E_{xc}[n]$, is not known in general. Using the following Euler-Lagrange equation based on the variational principle

$$\frac{\delta E[n(r)]}{\delta n(r)} + \mu \frac{\delta\left(N - \int n(r)dr\right)}{\delta n(r)} = 0 \qquad (16)$$

where μ is the Lagrange multiplier taking care of the particle number conservation, we get

$$\frac{\delta T[n(r)]}{\delta n(r)} + V_{ion}(r) + V_H(r) + \frac{\delta E_{xc}[n(r)]}{\delta n(r)} = \mu \qquad (17)$$

By splitting up T[n] into the kinetic energy of non-interacting electrons, T_0, plus a remainder, T_{xc}, which may be conveniently incorporated into the exchange-correlation energy (E_{xc}), Eq. 17 can be interpreted as the Euler-Lagrange equation of non-interacting electrons moving in an external effective potential given by

$$V_{eff}(r) = V_{ion}(r) + V_H(r) + V_{ex}(r) \tag{18}$$

where the Hartree potential is

$$V_H(r) = \int \frac{n(r')}{|r - r'|} dr' \tag{19}$$

and the exchange-correlation potential which contains all the many-body effects is

$$V_{ex}(r) = \frac{\delta E_{xc}[n(r)]}{\delta n(r)}. \tag{20}$$

Thus the interacting multi-electron system can be simulated by the non-interacting electrons under the influence of the effective potential $V_{eff}(\mathbf{r})$. Kohn and Sham [51] took advantage of the fact that the case of non-interacting electrons allows an exact computation of the particle density and kinetic energy as

$$n(r) = \sum_{i=1}^{N} \varphi_i^*(r)\varphi_i(r) \tag{21}$$

$$T_0[n(r)] = \sum_{i=1}^{N} \left(-\frac{1}{2}\right) \int \varphi_i^*(r)\nabla^2\varphi_i(r)dr \tag{22}$$

Writing the Euler-Lagrange equations in terms of the single-particle wave functions (φ_1) the variation principle finally leads to the effective single-electron equation, well-known as the Kohn-Sham (KS) equation:

$$\left\{-\tfrac{1}{2}\nabla^2 + V_{eff}(r)\right\}\varphi_i(r) = \varepsilon_i\varphi_i(r) \tag{23}$$

where the KS eigenvalues (ε_i) have, strictly speaking, no physical meaning as the energies of the single-particle electron states. The KS equation constitutes a self-consistent field problem, that is, the self-consistent solution (φ_i) can be obtained by iteratively solving the KS equation (Eq. 23) and the total electron density is determined from Eq. (21). The ground-state total energy is given by the sum of the KS eigenvalues minus the so-called double counting terms:

$$E = \sum_{i}^{N} \varepsilon_i - \frac{1}{2} \int \frac{n(r)n(r')}{|r - r'|} dr \; dr' + G_{xc}[n(r)] - \int V_{xc}(r)n(r)dr \qquad (24)$$

Use of the Born-Oppenheimer approximation is implicit for any many-body problem involving electrons and nuclei as it allows us to separate electronic and nuclear coordinates in many-body wave function. Because of the large difference between electronic and ionic masses, the nuclei can be treated as an adiabatic background for instantaneous motion of electrons. So with this adiabatic approximation the many-body problem is reduced to the solution of the dynamics of the electrons in some frozen-in configuration of the nuclei. However, the total energy calculations are still impossible without making further simplifications and approximations.

3.4
Approximations

3.4.1
Electron-Ion Interactions and Pseudopotentials

The interaction between the electrons in a molecule and the atomic nuclei is normally a very strongly attractive one and in general it would be necessary to solve for the energy levels of all electrons in the molecule. However for many purposes the problem can be simplified by separating the electrons into valence and core types. The valence electrons are those which form the molecular electronic charge distribution. The core electrons are those for which the energy levels are unperturbed relative to the free atom and therefore these electrons do not participate in intramolecular bonding. The core electrons screen the nuclear charge leaving a weak residual electrostatic potential which is experienced by the valence electrons. The electronic structure problem can in principle therefore be confined to valence electrons only.

The basic idea of the pseudopotential theory is to replace the strong electron-ion potential by a much weaker potential – a pseudopotential that can describe the salient features of the valence electrons which determine most physical properties of molecules to a much greater extent than the core electrons do. Within the pseudopotential approximation, the core electrons are totally ignored and only the behaviour of the valence electrons outside the core region is considered as important and is described as accurately as possible [54]. Thus the core electrons and the strong ionic potential are replaced by a much weaker pseudopotential which acts on the associated valence pseudo wave functions (φ^{ps}) rather than the real valence wave functions (φ^r). As shown in Fig. 7, the two types of potentials and wave functions are identical at and outside the core radius (r_c) so that the behaviour of the valence electrons in this region is well described by the pseudopotentials and pseudo wave functions. However, inside the core the rapid oscillations of the true valence wave functions are no longer present in the pseudo wave functions.

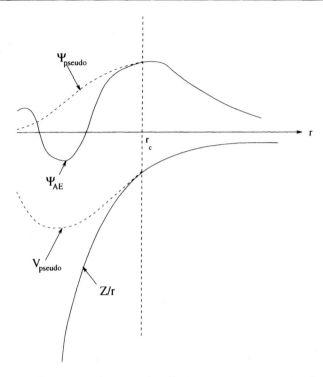

Fig. 7. Schematic illustration of the pseudo φ^{ps} (*dashed curves*) and real φ^r (*solid curves*) valence wave functions and associated potentials V^{ps} and Z/r as a function of r

It remains to construct an accurate ionic pseudopotential from first principles which will be appropriate for a variety of molecular environments. There have been some very recent developments in the construction of ab initio pseudopotentials and we will only discuss the fundamental strategies [55, 56].

Psueudopotentials should satisfy several basic requirements. For example, the pseudo (φ^{ps}) and real (φ^r) wave functions must be identical outside the core radius ($>r_c$), not only in their spatial dependence but also in their absolute magnitudes such that two wave functions generate identical charge densities. The equality of the two types of wave functions outside the core radius in this context is guaranteed by imposing the following constraint:

$$\int_0^{r_c} |\varphi^r(r)|^2 r^2 d^3r = \int_0^{r_c} |\varphi^{ps}(r)|^2 r^2 d^3r \tag{25}$$

which ensures that the charge enclosed within the core radius for two wave functions must be equal. If the pseudopotential satisfies this condition, it is commonly referred to as a *norm-conserving* pseudopotential.

3.4.2 Electron-Electron Interactions

The interactions between electrons are inherently many-body forces. There are several methods in common use today which try to incorporate some, or all, of the many-body quantum mechanical effects. An important term is that of electronic exchange [57, 58]. Mathematically, when two particles in the many-body wavefunction are "exchanged" the wavefunction changes sign:

$$\Psi(r_1, \ldots, r_j, \ldots, r_k, \ldots) = -\Psi(r_1, \ldots, r_k, \ldots, r_j, \ldots). \tag{26}$$

To ensure this, the-many-body wavefunction can be written as a "Slater determinant" of one particle wavefunctions – this is the Hartree Fock method. The drawbacks of this method are that it is computationally demanding and does not include the many-body correlation effects.

An alternative approach is to try to deduce an approximate effective potential for the combined effects of exchange and correlation. This is the main principle of DFT. To solve the KS equation (Eq. 23) requires the precise knowledge of the exchange-correlation potential V_{xc}. It is the local density approximation (LDA) which describes the exchange-correlation energy by taking the exchange-correlation energy per electron at a point r in the electron gas, $E_{xc}(r)$, to be equal to the exchange-correlation energy per electron in a homogeneous electron gas that has the same density as the electron gas at point r. The local representation of the exchange-correlation energy functional and corresponding potential are

$$E_{xc}^{LDA}[n(r)] \equiv \int n(r) E_{xc}(n(r)) dr \tag{27}$$

$$E_{xc}^{LDA}(r) = \frac{\delta E_{xc}^{LDA}[n(r)]}{\delta n(r)} = \frac{\partial [n(r) E_{xc}(n(r))]}{\partial n(r)} \tag{28}$$

The exchange-correlation energy density can be split into two parts: exchange component $E_x(n)$ and correlation component $e_E(n)$. The explicit expression for the exchange component is known from Hartree-Fock theory but the correlation component is known only numerically. Several parametrisations exist for the exchange-correlation energy and potential of a homogeneous gas system which can be used for the LDA calculations within DFT.

The local density approximation is highly successful and has been used in density functional calculations for many years now. There were several difficulties in implementing better approximations, but in 1991 Perdew et al. successfully parametrised a potential known as the generalised gradient approximation (GGA) which expresses the exchange and correlation potential as a function of both the local density and its gradient:

$$E_{xc}^{GGA} \sim E_{xc}^{GGA}[n(r), \nabla n(r)]. \tag{29}$$

which has recently been shown to improve on the LDA in several classes of material, but especially in molecular and hydrogen bonded systems. Most of the ab initio calculations described in this article therefore use the GGA.

4
Intramolecular Bonding and Electronic Properties of Liquid Crystal Molecules

4.1
Molecular Electronic Structure

4.1.1
Basis Sets and Boundary Conditions

For practical implementations it is necessary to represent the molecular electronic wave functions as a linear combination of some convenient set of basis functions. In principle any choice of basis set is permissible although the basis set must span any electronic configuration of the molecule. This implies that the basis must form a complete set.

Bloch's theorem states that in a periodic solid each electronic wave function can be expressed as the product of a wave-like component (with wave vector $\mathbf{k}$) and a cell-periodic component $f_i(\mathbf{r})$:

$$\varphi_i(r) = \exp[I\mathbf{k} \cdot r] \, f_i(r) = \sum_G c_{i,\mathbf{k}+G} \, \exp[i(\mathbf{k} + G) \cdot r] \tag{30}$$

where the cell periodic function $f_i(\mathbf{r})$ is expanded using a basis set consisting of a discrete set of plane waves corresponding to the reciprocal lattice vectors G of the crystal. Thus one of the simplest approaches is to expand the molecular electronic wavefunction in a series of plane waves. This amounts to making a Fourier expansion of the charge density. The wavefunctions are expanded up to some energy cut-off $E_{cut-off}$ to truncate the basis set such that the coefficients $c_{j,\mathbf{k}+G}$ in Eq. (30) for the plane waves with kinetic energy $(\hbar^2/2m)|\mathbf{k} + G|^2$ higher than the cut-off energy are very small compared to those with lower kinetic energies. The magnitude of the error due to the truncation of the plane-wave basis set can always be reduced by increasing the value of the cut-off energy.

The KS equation (Eq. 23) when expressed in terms of a plane-wave basis set takes a very simple form:

$$\sum_G \left[\frac{\hbar^2}{2m} |\mathbf{k} + G|^2 \delta_{GG'} + V_{\text{eff}}(G - G') \right] c_{i,\mathbf{k}+G'} = \varepsilon_i c_{i,\mathbf{k}+G} \tag{31}$$

$$H_{\mathbf{k}-G,\mathbf{k}-G'} \, c_{i,\mathbf{k}+G'} = \varepsilon_i c_{i,\mathbf{k}+G} \tag{32}$$

where $V_{eff}(G-G')$ is the Fourier transform of the effective potential in Eq. (23). The coefficients enter a variational calculation for the minimum energy.

There are several advantages associated with the plane wave approach. For example:

- the method is conceptually simple;
- completeness is guaranteed and convergence toward completeness can easily be tested;
- the same basis set can be used for all atomic species in the molecule provided that psuedopotentials exist;
- force calculations do not require corrections.

The minimisation problem can be distributed over the nodes of a parallel computer, thereby allowing the method to be applied to very large molecules.

Since plane waves are delocalised and of infinite spatial extent, it is natural to perform these calculations in a periodic environment and periodic boundary conditions can be used to enforce this periodicity. Periodic boundary conditions for an isolated molecule are shown schematically in Fig. 8. The molecular problem then becomes formally equivalent to an electronic structure calculation for a periodic solid consisting of one molecule per unit cell. In the limit of large separation between molecules, the molecular electronic structure of the isolated gas phase molecule is obtained accurately.

4.1.2
Charge Distribution

The molecular electronic charge density can be reconstructed directly from the individual wavefunctions for occupied levels according to

$$n(\vec{r}) = \sum_i |\varphi_i(r)|^2 \qquad (33)$$

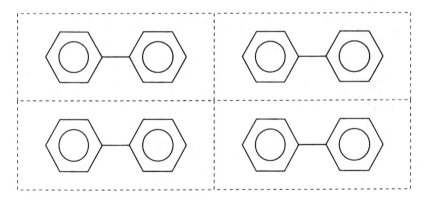

Fig. 8. Illustration of the use of periodic boundary conditions in the determination of molecular electronic structure. The unit cell is shown by the *dashed line*. As the unit cell size is increased the calculated properties converge toward those of the isolated molecule

where the sum runs over all occupied levels. The implication of this is that the molecular electronic charge distribution can be examined directly. In Fig. 9 we show the molecular electronic charge distribution for 5CB (Fig. 9a) and the ferroelectric liquid crystal DOBAMBC (Fig. 9b). From the charge distribution a very wide variety of molecular properties can be deduced. We discuss some examples in the next section.

The use of periodic boundary conditions also allows for an efficient evaluation of the ion-ion interaction. Ewald developed a method to compute the Coulomb energy associated with long range ion-ion interactions in solids. The Coulomb energy due to interactions between an ion at position R_2 and an array of ions positioned at R_{1+l} is given by

$$E_{\text{ion}} = \frac{1}{2} \sum_{ij} \sum_{l} \frac{Z_i Z_j e^2}{|R_1 + l - R_2|} \tag{34}$$

where l is the lattice vectors, Z_i and Z_j are the valences of ions i and j. Since the interaction is long ranged in both the real and reciprocal spaces, it is very

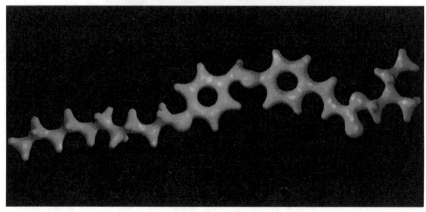

Fig. 9a,b. The molecular electronic charge distributions for: **a** the nematogen 5CB; **b** the ferroelectric DOBABMC, as constructed directly from the molecular electronic wavefunctions using $n(\vec{r}) = \sum_l |\varphi_i(r)|^2$.

difficult to calculate E_{ion}. The Ewald method allows a replacement of the infinite Coulomb summation by two rapidly convergent summations, one over lattice vectors and the other over reciprocal lattice vectors (**G**). The expression for the Ewald energy of the ionic system is

$$E_{ion} = \frac{1}{2}\sum_{ij} Z_i Z_j e^2 \sum_{l} \frac{erfc(\eta)|R_1 + l - R_2|)}{|R_1 + l - R_2|} - \frac{2\eta}{\sqrt{\rho}}\delta_{ij}$$
$$+ \frac{4\pi}{\Omega}\sum_{G\neq 0}\frac{1}{G^2}\exp\left(-\frac{|G|^2}{4\eta^2}\right)\cos[(R_1 - R_2)\cdot G] - \frac{\pi}{\eta^2\Omega}$$

(35)

where *erfc* is the complementary error function, η is the value used to divide the sum into two sums and Ω is the volume of the unit cell.

4.2
Electric Multipole Moments

4.2.1
Molecular Dipoles

Liquid crystal molecules are usually of very low symmetry and most possess a non-vanishing permanent dipole moment which couples to applied fields and is principally responsible for field-induced molecular reorientation. In addition to this, dipolar forces are usually of very long range and have been shown to be responsible for intermolecular orientational correlations in condensed phases. Molecular dipoles therefore represent a clear bridge between intramolecular electronic structure and long-range intermolecular forces. The molecular dipole moment is therefore one of the most important properties which can be determined directly from knowledge of the ground state charge distribution and the relaxed molecular structure. The molecular dipole itself is a measure of the difference between the centres of positive and negative charge on a molecule. It can therefore be determined from electronic structure calculations if the ionic positions and electronic charge distribution are known. In the case of 5CB density functional plane wave techniques have shown that the molecular dipole can be estimated accurately from the calculated moments of suitably chosen fragments [59]. The molecular dipole moments of mesogenic fragments such as fluorinated biphenyl have also been reported [59].

4.2.2
Electric Quadrupoles

In principle, the electronic quadrupole can also be extracted from the calculated valence charge distribution. This is a tensor quantity and its components are defined by the matrix

$$\frac{1}{2}\begin{pmatrix} \sum q(3x_i^2 - r_i^2) & 3\sum q_i x_i y_i & 3\sum q_i x_i z_i \\ 3\sum q_i y_i x_i & \sum q_i(3y_i^2 - r_i^2) & 3\sum q_i y_i z_i \\ 3\sum q_i z_i x_i & 3\sum q_i z_i y_i & \sum q_i(3z_i^2 - r_i^2) \end{pmatrix} \tag{36}$$

where q_i is the electronic or ionic charge, x_i, y_i and z_i are Cartesian components and r is the distance from an origin. Diagonalisation leads to the quadrupole in the molecular principle axis.

Currently, the effect of the molecular quadrupoles on liquid crystal properties is not clearly understood though there is some evidence to suggest that they influence surface properties and phase stability [28, 60].

4.2.3
Polarisability and Its Anisotropy

Polarisability relates the dipole moment of a molecule linearly to an applied electric field

$$\mathbf{D} - \mathbf{D}_0 = \alpha \mathbf{E} + \beta \mathbf{E}^2 + \dots \tag{37}$$

where $\mathbf{D}$ is the dipole moment of the molecule, $\mathbf{D}_0$ is the permanent static dipole of the molecule, $\mathbf{E}$ is the applied electric field and α is the first order molecular polarisability which is a tensor quantity. These can be straightforwardly calculated within an ab initio method by adding an additional potential, V to the Hamiltonian, where

$$\mathbf{E} = -\nabla V \tag{38}$$

A sequence of calculations can be performed with various applied electric fields in which the dipole moment of the molecule is evaluated, as described above. The 3×3 polarisability tensor, α_{ij} can therefore be constructed.

For the linear response of the dipole to an electric field, this calculation is fairly straightforward. However, the dipole can also be calculated for a range of magnitudes of applied field to obtain higher order hyperpolarisabilities such as β.

5
Molecular Structure and Flexibility

5.1
Overview

Although liquid crystalline phases can be formed from entirely rigid units such as rods and ellipsoids, all liquid crystal molecules are partially flexible objects which can change their shape in a variety of ways. The extension of rigid-rod theories to account for flexibility is complex and has been the object of considerable effort for several years. It is now clear that simply relaxing the assumption of molecular rigidity is expected to have profound effects on phase behaviour.

Analytical approaches to understanding the effect of molecular flexibility on orientational order have concentrated on both the isotropic-nematic and the nematic-smectic transition [61, 62] and mean field theory has shown that cholesteric pitch appears not to depend on the flexibility of the molecule [63].

Computational studies of non-rigid liquid crystals vary widely in their degree of specification to real systems. Molecular models have ranged from hard spheres joined by flexible links [64] to near-atomistic level treatments of mesogens [65, 66]. At a molecular level, flexibility is a measure of how easily the electron distribution responsible for the molecular bonding can be distorted from its equilibrium configuration. The central quantity is the torsional potential which represents the energy cost associated with twisting various segments of the molecule relative to others. In this section, the molecular flexibility of liquid crystal molecules and related fragments is examined on the basis of electronic structure with emphasis on the shape dependence of properties known to be important in liquid crystal phase formation.

5.2
Geometry Optimisation

In simple molecules having no conformational flexibility, there are still geometrical degrees of freedom arising from bond length and bond angle variations. The ionic positions can be optimised according to the corresponding equations of motion if the forces acting on the ions are known. However, care must be taken in calculating these forces since the minimum of the KS energy functional is physically meaningful only within the Born-Oppenheimer (BO) approximation. Any change of the ionic configuration must be accompanied by an instantaneous redistribution of the electrons so the electronic wave functions must change accordingly to maintain the system on the BO surface in the multidimensional space of $(E, \varphi, \mathbf{R}_I)$. This inevitable change in the wave functions with $\mathbf{R}_I$ affects the minimum of the KS energy functional and, therefore must be taken into account in calculating the forces on ions. The force acting on an ion at position $\mathbf{R}_I$ can, in general, be written as

$$\mathbf{F}_I = -\frac{dE}{d\mathbf{R}_I} = -\frac{\partial E}{\partial \mathbf{R}_I} - \sum_i \frac{\partial E}{\partial \varphi_i} \frac{\partial \varphi_i}{\partial \mathbf{R}_I} - \sum_i \frac{\partial E}{\partial \varphi_i^*} \frac{\partial \varphi_i^*}{\partial \mathbf{R}_I} \quad (39)$$

According to the Hellmann-Feynman theorem [67], when the wave function is an eigenstate of the Hamiltonian, the last two terms in Eq. (39) are individually zero since they are simply related to the derivative of the normalisation constant $\langle \varphi_i | \varphi_i \rangle$ with respect to $\mathbf{R}_I$. The theorem simplifies the calculation of the physical forces on the ions, although the electronic wave functions must be eigenstates of the KS Hamiltonian. Since the forces have a first order error with respect to deviations in the wave functions from the ground state, the forces should be calculated only when the electron configuration is sufficiently near its ground state, i.e. the electrons are fully relaxed to the equilibrium configuration, otherwise the error in forces may cause the ionic configuration to drift away from the local minimum. Moreover, a fine tolerance of convergence of the wavefunctions is necessary in each step of the ionic relaxation to prevent instabilities.

The Hellmann-Feynman forces are also sensitive to the effect of moving ions on the basis set (φ_j) of the electronic wave function ($\varphi_i = \Sigma_j c_{ij} \phi_j$). This contribution to the force from the variation of the basis set with the ionic position ($d\phi_j / d\mathbf{R}_I \neq 0$) is known as the Pulay force [68]. However, for the plane wave basis set $d\phi_j/d\mathbf{R}_I$ is zero and hence the Pulay force is zero. Thus the advantage of using a plane wave basis set over a localised one is that the error in the Hellmann-Feynman force is directly related to the extent to which the electronic configuration is converged with respect to its ground state, and may therefore be reduced to an arbitrarily small value.

The molecular dynamics associated with the ionic relaxation can be performed using a minimisation scheme. Once the forces on the ions are calculated by converging electronic wave functions to the ground state, the ionic positions can be optimised according to the calculated forces. Each time the ions are allowed to move in the direction of the calculated Hellmann-Feynman forces, the electrons must be brought to the ground state of the new ionic configuration. This iterative process is continued until the ionic positions relax finally to the equilibrium configuration with zero forces acting on them and with minimum energy for a given unit cell.

Recent applications of this method have shown that isolated benzene retains perfect six-fold symmetry to within numerical accuracy of the calculation [69] whereas in 5CB subtle symmetry breaking in the constituent phenyl rings occurs by contraction of the C—C bonds along the long axis of the molecule [69].

5.3
Torsional Potentials of Common Liquid Crystals and Fragments

Biphenyl is a prototypical semi-rigid fragment of many liquid crystal molecules and its structure and flexibility have been studied using a variety of experimental and theoretical approaches. Biphenyl affords a simple example of a flexible system having a single torsional angle, the torsional potential for which is determined by two competing interactions. This competition arises from the fact that the π conjugation between the phenyl rings favours the planar orientation whereas the steric repulsion between the hydrogen atoms favours the tilted structure. Gas phase electron diffraction experiments have shown that the equilibrium ring angle is approximately $44.4° \pm 1.4°$ [70] though this value seems to be sensitive to condensed phase environments.

A variety of molecular orbital strategies of varying levels of sophistication have been applied to the biphenyl case. Early applications of ab initio methods to biphenyl were reported by McKinney et al. (HF/STO approach) and Almöf using "double χ" Gaussian basis sets. The equilibrium ring angle was found to be $42°$ and $32°$, respectively, using fixed bond angles and lengths. Subsequent studies have gone further, implementing full geometry relaxation [71, 72] using basis sets of the type STO/3G [71] and 6-31G [72]. The calculated torsional angle was found to depend sensitively on the type of basis set used with the 6-31G choice, giving good agreement with electron diffraction results. The full rotational potential of biphenyl was calculated at the HF/6-31G* level [73] using molecular orbital methods. Geometry was obtained at the Hartree Fock level and and Moller-Plesset perturbation methods [74] were used to correct for correlation effects. DFT results are shown in Fig. 10.

Studies of molecular flexibility for single nematogens in gas phase have now been performed for 5CB and related fragments using DFT plane wave methods [75]. The conformational landscape showing the dependence of total energy on ring angle and all trans tail angle is shown in Fig. 11. The lowest energy conformation corresponds to the ring angle at 31° and the tail angle at 90°. However, this minimum is quite shallow especially with respect to tail orientation.

5.4
Shape-Dependent Molecular Properties

In addition to the geometric, steric effects which flexibility directly modifies, all intramolecular electronic properties of a flexible molecule are also coupled to molecular structure. In a real material, such properties as molecular dipole,

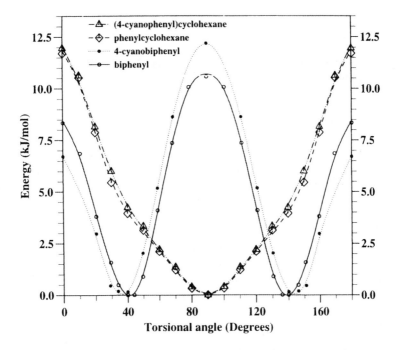

Fig. 10.   The torsional potential for biphenyl as a function of inter-ring angle

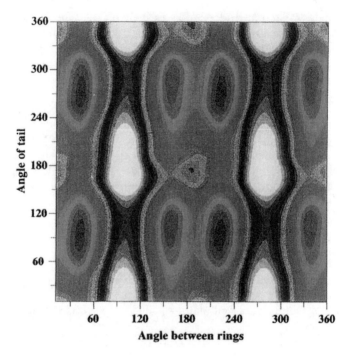

Fig. 11. Torsional inter-ring potential landscape for the nematogen 5CB

quadrupole and polarisability must be taken as statistically weighted averages over the range of conformers.

Moreover, intermolecular interactions which arise from them must also take account of the coupling to structural changes. In ferroelectrics, measurements are reported of the ferroelectric polarisation P_S induced in a non-chiral smectic C phase by a variety of chiral dopants having different molecular structural features. Using molecular calculations of contributing dipole moments, ferroelectric order parameters are determined from the experimental results. The relationships between the P_S and various other molecular properties are discussed, and it is shown that restricted rotation of the molecule due to its shape and internal energy barriers to rotation can result in relatively high values of P_S. In contrast dipolar groups flexibly attached to a chiral centre may have their contribution to P_S greatly reduced through internal rotation [76].

In some molecules the molecular dipole is simply dominated by a single polar bond. In sufficiently complex molecules there may be several polar bonds of differing strength. In this case the molecular dipole is determined by their relative orientation. Ab initio studies of 2-2′-difluoro biphenyl have revealed a strong shape dependence of the molecular dipole as a function of inter-ring angle. This is illustrated in Fig. 12.

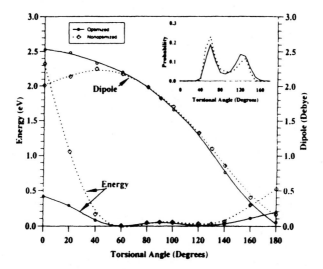

Fig. 12. Torsional potential and molecular dipole for 22-difluoro biphenyl

5.5
Transferability of Torsional Potentials

5.5.1
Influence of the Intramolecular Environment

So far, current strategies for extracting reliably the properties of isolated liquid crystal molecules have been outlined. From this it is possible to extract useful intramolecular potentials governing, for example, molecular flexibility. It is clear however that molecular structure and flexibility may be modified by chemical substituents and by the condensed phase environment. In view of this, studies of the transferability of intramolecular potentials are particularly valuable. The aim of such investigations is to determine the extent to which ab initio derived potentials can be used in environments which differ from those in which they have been calculated. Further to this is the related question of whether potentials derived from small fragments can be combined to give an accurate representation of the intramolecular properties of a much larger system. Very few such transferability studies have been performed for mesogens [77] with the emphasis of these being on inter-ring torsions in substituted biphenyls and phenylcyclohexane. In these cases it is found that the transferability of the torsional potential depends on the degree of electron delocalisation (conjugation) in the mesogenic core with transferability decreasing with increasing conjugation.

6
Vibrational Properties

6.1
Importance of Vibrational Properties of Liquid Crystals

In addition to molecular electronic structure and conformational flexibility, liquid crystal molecules also undergo complex internal vibrations. The vibrational frequencies of molecules are very important properties for two main reasons: first, they provide a direct measure of the bond strength and second, they are fairly easy to measure experimentally. Moreover, the calculation of accurate vibrational frequencies is a far more stringent test of the predictive power of simulations than is the calculation of equilibrium geometries. This is because the mode frequency (bond strength) is determined by the curvature of the interatomic potential whereas the equilibrium bond lengths and angles are determined only by the minimum of the potential [78]. In view of the importance of molecular flexibility in liquid crystal systems, it is essential to be able to describe distorted configurations of molecules with no loss of accuracy and the comparison of observed and calculated vibrational frequencies provides a measure of this capability.

In the case of liquid crystals in particular, vibrational properties reflect very directly the complex hierarchy of the structure and bonding problem in these materials. For example, in a single mesogenic molecule vibrational frequencies range from about 10 cm^{-1} to over 3000 cm^{-1} which arise from the very wide range of force constants present [79].

6.2
Experimental Studies of Molecular Vibrations in Liquid Crystals

Experimental studies of liquid crystals have been used for many years to probe the dynamics of these complex molecules [12]. These experiments are usually divided into high and low-frequency spectral regions [80]. This distinction is very important in the study of liquid crystalline phases because, in principle, it can discriminate between inter- and intramolecular dynamics. For many organic materials vibrations above about 150 cm^{-1} are traditionally assigned to internal vibrations and those below this value to so-called "lattice modes". However, the distinction is not absolute and coupling between inter- and intramolecular modes is possible.

Normal vibrational spectroscopy generates information about the molecular frequency of vibration, the intensity of the spectral line and the shape of the associated band. The first of these is related to the strength of the molecular bonds and is the main concern of this section. The intensity of the band is related to the degree to which the polarisability is modulated during the vibration and the band shape provides information about molecular reorientational motion.

One of the most common uses of vibrational spectroscopy is as a molecular "fingerprinting" tool whereby spectral features are assigned to the presence of particular fragments in molecules. These assignments are, however, only approximate and in complex molecules such as liquid crystals significant perturbation of the molecular mode frequencies relative to the isolated fragments can occur. An understanding of the origin of vibrational bands in mesophases is important in studies of order parameter, phase transitions and molecular reorientational correlation functions.

The study of intermolecular modes in organic molecules generally and liquid crystals in particular is less straightforward from an experimental point of view. Such modes exist close to the Rayleigh line and very good stray light rejection is needed for their observation. One would expect that intermolecular potentials responsible for any low-frequency features would arise from electrostatic potentials which scale with distance as $(1/r^n)$ where n is greater than 3. Therefore the resulting force constants are rapidly decreasing functions of distance. In the next section we outline a scheme for calculating the vibrational frequencies of molecules from first principles.

6.3
Calculation of Mode Frequencies of Large Molecules from First Principles

The starting point for the determination of mode frequencies is the harmonic approximation. Here we assume that we begin with an equilibrium geometry and investigate the restoring forces generated for small displacements from equilibrium.

Frequencies ω and polarisation vectors $\mathbf{u}_\kappa$ are determined by the diagonalisation of the dynamical matrix

$$D^{\alpha\beta}_{\kappa\kappa'} = \frac{1}{\sqrt{m_\kappa m_{\kappa'}}} \, \Phi^{\alpha\beta}_{\kappa\kappa'} \tag{40}$$

which satisfies the eigenvalue problem

$$\sum_{\kappa'} D_{\kappa\kappa'} \, u_{\kappa'} = -\omega^2 \mathbf{u}_\kappa \tag{41}$$

with eigenvalues giving the frequencies and eigenvectors giving the polarisations of the vibrational modes.

The molecule contains atoms with mass m_κ at positions r_κ and α and β in the above equations correspond to Cartesian axes (x, y, z).

The real space interatomic force constants $\Phi^{\alpha\beta}_{\kappa\kappa'}$ can be derived from the total energy E of the system

$$\Phi^{\alpha\beta}_{\kappa\kappa'} = \frac{\partial^2 E}{\partial u^\alpha_\kappa \partial u^\beta_{\kappa'}} \tag{42}$$

or, alternatively, from forces acting on the atoms

$$\Phi^{\alpha\beta}_{\kappa\kappa'} = \frac{\partial F^\alpha_\kappa}{\partial u^\beta_{\kappa'}} \tag{43}$$

where F_{κ}^{α} is the force experienced by the atom κ along the α direction due to displacement $u_{\kappa'}^{\beta}$ of the κ' atom along the β direction.

Subsequent diagonalisation of the dynamical matrix constructed from first-principles simply yields the required mode frequencies and eigenvectors. It is important that the atomic displacements are to be made small enough to ensure the harmonicity. The effect of anharmonicity can be greatly reduced by making positive and negative displacements about the equilibrium positions and averaging the corresponding restoring forces. The size of the harmonic displacements is generally taken to be of the order of 0.005 in fractional coordinates.

Benzene has often been used as a test system for vibrational calculations using a variety of different electronic structure algorithms. The molecule exhibits regular hexagonal planar symmetry with six carbon atoms joined by σ bonds and six remaining p-orbitals which overlap to form a delocalised π electron over all six carbon atoms. Table 1 shows comparisons of several different methods for benzene.

6.4
Low Frequency Dynamics in Liquid Crystals

The low frequency Raman spectrum of 5CB in the nematic phase and several solid polymorphs is shown in Fig. 13.

It is clear that the nematic phase exhibits a featureless Rayleigh wing and that several distinct solid phases can be formed depending on cooling rate [79]. This includes an apparently glassy phase. The vertical tick marks indicate the calculated frequencies of vibrational modes as obtained from density functional methods.

It is also evident that over the spectral region 50 cm^{-1} to 200 cm^{-1} considerable overlap exists between intramolecular torsional modes and intermolecular phonon-like excitations [79]. It therefore appears that low-lying vibrational excitations in 5CB and related liquid crystals do not show a clear separation between inter- and intramolecular character. However, below 50 cm^{-1} there are no calculated single molecule torsional modes and the few observed spectral features in this region are expected to represent progressively better defined intermolecular motion similar to lattice modes of conventional molecular or layered solids [81]. For a molecule as complex as 5CB, it is very difficult to assign modes to individual displacement patterns without ambiguity. The power of electronic structure methods allows for a direct calculation of normal mode displacement patterns. For example a selection of eigenvectors for 5CB is shown in Fig. 14.

In 8CB, continued cooling into the smectic phase reveals the appearance of a broad ultra-low-frequency feature centred at around 10 cm^{-1} where no other modes are seen. This is shown in Fig. 15. This feature appears to be unique to the smectic phase and has been tentatively attributed to intermolecular dipolar coupling across smectic layers [79]. In principle this should be a generic feature of smectics but it will be necessary to explore this issue through extensive computer simulations using realistic, shape-dependent potentials for

Table 1. Experimentally observed frequencies of benzene and also the frequencies found from DFT-plane wave calculations (the units are in cm^{-1}) [69]. Also shown, are results from several different commonly used quantum chemistry methods and basis sets. This fully ab initio structural calculation gives C—C bonds of length 1.396 Å and C—H of length 1.089 Å (as compared to experimental bond-lengths of 1.393–1.398 Å and 1.081–1.090 Å respectively). No symmetry of any kind was imposed on the structure (i.e. all 12 atoms were treated independently. The sixfold and planar symmetries exist to within a tolerance of 0.001 Å on positions. Normal mode frequencies are calculated as described in the main text. Agreement with experiment to better than 2% is obtained in most cases. Vibrational eigenvectors from which mode symmetries were assigned as shown. Frequencies as calculated by more conventional methods compare equally well, if not better in most cases. The main difference is that in the DFT-PW method, the same delocalised basis set is used and convergence towards completeness is easily tested. Results from the most commonly used HF-MP2 method are shown with progressive 'improvements' to localised basis sets. Note that, for example, the B_{2u} mode at around 1300 cm^{-1} actually gets worse with increased sophistication. This indicates that, even for the most 'reliable' localised basis sets, convergence is not being achieved. Therefore, unlike the plane wave basis set, the localised sets cannot be reliably used to determine molecular properties of as yet unsynthesised molecules where no a priori knowledge of the nature of bonding (and hence the type of basis set) is available.

Sym.	Obs. (cm^{-1})	DFT PW	MP2 (6-31G)	MP2 (6-31G(dp))	MP2 (TZ2P+f)	CCSD (cc-VTZ)	DFT (6-31G*)
A_{1g}	994	1006	1000	1015	1018	1012	1051
A_{1g}	3191	3260	3213	3240	3242	3228	3114
A_{2g}	1367	1351	1412	1367	1374	1391	1318
E_{2g}	608	597	632	610	608	613	606
E_{2g}	3174	3208	3179	3215	3217	3204	3096
E_{2g}	1607	1607	1626	1645	1637	1672	1645
E_{2g}	1178	1181	1234	1199	1195	1207	1157
B_{1u}	1010	1008	1046	1009	1039	1025	1017
B_{1u}	3174	3199	3173	3204	3218	3189	3092
B_{2u}	1309	1339	1377	1451	1461	1304	1411
B_{2u}	1150	1120	1229	1173	1178	1166	1129
E_{1u}	1038	1042	1073	1063	1074	1071	1043
E_{1u}	1494	1476	1527	1509	1515	1528	1475
E_{1u}	3181	3224	3203	3231	3238	3221	3108
B_{2g}	990	993	–	–	–	–	–
E_{2u}	967	962	–	–	–	–	–
B_{2g}	707	722	–	–	–	–	–
E_{2u}	398	399	–	–	–	–	–
A_{2u}	674	667	–	–	–	–	–
E_{1g}	847	833	–	–	–	–	–

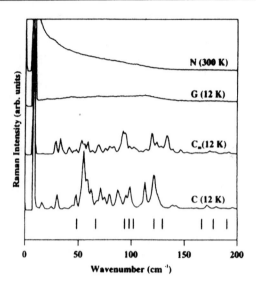

Fig. 13. The low-frequency vibrational mode spectrum for 5CB in several phases. The calculated mode frequencies are shown as *tick marks along the bottom*

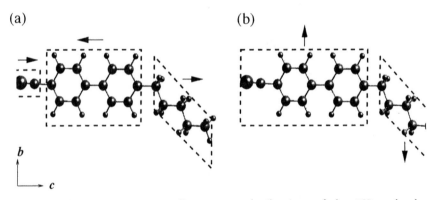

Fig. 14a,b. Eigenvectors corresponding to several vibrations of the 5CB molecule as calculated from first principles. *Arrows* denote direction of atomic vibrational motion

the molecules which allow for quantitative estimates to be made as to the strength of coupling which is expected.

7
Forward Look

In this article, methods for extracting ground state structural, dynamical and molecular electronic properties for liquid crystals have been outlined. It is clear that these methods have been applied only to a small number of systems

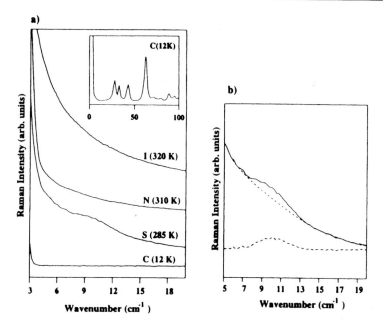

Fig. 15. The low-frequency vibrational mode spectrum for 8CB in its smectic phase. The broad and weak spectral feature at about 8 cm^{-1} has been attributed to intermolecular forces extending across smectic layers

and there is much scope for future work. Certainly the most obvious next step is to incorporate them, where possible, into studies of condensed phase behaviour. Combining calculations of shape-dependent electronic dipoles and polarisabilities with information on long range orientational order could be used to extract very accurate data on dielectric anisotropy. Density functional studies of the interactions between two or more mesogenic molecules can give insight into the influence of short-ranged structural correlations on molecular properties. Such an approach would make it possible to begin to build rudimentary bridges between isolated molecule structure and condensed phase properties, thereby addressing the single-molecule structure property problem. Alternatively, and as alluded to earlier, one could consider using the results of ab initio quantum mechanical methods for parametrisation of potentials for use in classical molecular dynamics or Monte Carlo simulations of condensed phases which would be too large to treat using entirely parameter free methods [82]. The initial steps towards this goal have already been taken and the first simulations of liquid crystalline phases and properties have been performed using atomic-level detail.

The interaction of complex liquid crystal molecules with realistic surfaces is an area which is currently unexplored using electronic structure methods though, as stated earlier, the problem of surface-induced control of molecular orientation remains at the forefront of liquid crystal device technology. This problem is currently at the limits of practical capability of the most powerful computer systems. However treatment of a single mesogenic molecule on a silicon surface is probably not out of reach even at present. The sort of physical/chemical questions such an investigation might address include the following. (1) What is the chemical and electronic nature of the interactions between liquid crystal molecules and realistic surfaces? (2) How strong is the molecule to surface bond? (3) How free is the molecule to rotate in the plane of the surface or to tilt at an angle? (4) How are the properties of the molecules modified by the presence of the surface? (5) How might the properties of molecules or surfaces be altered to improve control over alignment?

In addition to all these considerations and possible routes for further work, there is also the salient point that first principles methods are, in principle, applicable to hypothetical and as yet unsynthesised systems. The implication of this is that the combination of such first principles techniques with state of the art molecular synthesis and characterisation technology holds substantial promise for solutions to structure property problems.

Acknowledgements. JC acknowledges support from the Royal Society of Edinburgh, the Royal Society of London and the Defence Research Agency, Malvern. The authors are also grateful to GJ Ackland, CJ Adam, DJ Cleaver and MR Wilson for helpful discussions and to the Edinburgh Parallel Computing Centre.

8
References

1. See Chapters on TGB phases and cholesteric liquid crystals in this volume
2. Allen MP, Evans GT, Frenkel D, Mulder BM (1993) Adv Chem Phys 86. Wiley
3. DeGennes PG, Prost J (1993) The physics of liquid crystals. Oxford University Press, Oxford
4. Allen MP, Tildesley DJ (1987) Computer simulations and liquids. Oxford University Press, Oxford
5. Onsager Ann L (1949) NY Acad Sci 51: 627
6. Gay GB, Berne BJ (1981) J Chem Phys 56: 4213
7. Luckhurst GR, Simmonds PSJ (1993) Mol Phys 80: 233
8. Kaszyinski P, Huang JP, Jenkins GS, Bairamov KA, Lipiak D (1995) Mol Cryst Liq Cryst 260: 315
9. Mehl GH, Goodby JW (1996) Chem Berichte 129: 521
10. Hanemann T, Haase W, Svoboda I, Fuess H (1995) Liq Cryst 19: 699
11. Emsley JW, Deluca G, Celebre G, Longeri M (1996) Liq Cryst 20: 569
12. Bulkin BJ (1974) Advances in liquid crystals. Academic Press, New York
13. Gray GW (1979) Liquid crystals and molecular structure. In: Molecular physics of liquid crystals. Academic Press, New York
14. Rettig (1985) Thesis, Halle/Salle
15. Marcelis ATM, Koudijs A, Sudholter EJR (1995) Liq Cryst 18: 843
16. Myrvold BO (1989) Liq Cryst 5: 1139
17. Jadzyn J, Czechowski G, Shonova NT (1988) Liq Cryst 3: 1637
18. Doane JW, Parker RS, Cvikl B, Johnson DL, Fishel DL (1972) Phys Rev Lett 28: 1694
19. Wulf A (1975) Phys Rev A 11: 365
20. See the article by N. Clark in this volume
21. Gray GW, Hird M, Toyne KJ (1991) Mol Cryst and Liq Cryst 204: 43
22. Khoo IC, Wu ST (1993) Optics and nonlinear optics of liquid crystals, 1st edn. World Scientific, Singapore

23. Maier W, Saupe A (1960) Z Naturforsch Teil A15: 287
24. Maier W, Saupe A (1958) Z Naturforsch Teil A13: 564
25. Maier W, Saupe A (1959) Z Naturforsch Teil A14: 882
26. Velasco E, Mederos L, Sluckin TJ (1996) Liq Cryst 20: 399
27. McMillan WL (1973) Phys Rev A 8: 1921
28. Goosens WJA (1987) Europhysics Lett 3: 341
29. Wurflinger A, Urban S (1997) Adv Chem Phys 98: 143
30. Blinov LM, Chigrinov VG (1996) Electro-optical effects in liquid crystals. Springer, Berlin Heidelberg New York
31. Vanakaras AG, Photinos DJ (1995) Mol Phys 85: 1089
32. GilVillegas A, McGrother SC, Jackson G (1997) Chem Phys Lett 269: 441
33. Derzhanski AI, Petrov AG (1979) Acta Phys Polonica A 55: 747
34. Meyer RB (1969) Phys Rev Lett 22: 918
35. Schmidt D, Schadt M, Helfrich W (1972) Z Naturforsch 271: 277
36. Gelbart WM, Ben-Shaul A (1982) J Chem Phys 77: 916
37. Singh Y, Singh U (1989) Phys Rev A 39: 4254
38. Osipov MA, Terentjev EM (1989) Z Naturforch Teil A44: 785
39. Archer LA, Larson RJ (1995) Chem Phys 103: 3108
40. Swanson BD, Sorensen LB (1995) Phys Rev Lett 75: 3293
41. Roos BO, Andersson K, Fulscher MP (1992) Chem Phys Letters 192: 5
42. Wu ST (1991) J Appl Phys 69: 2080
43. Crawford GP, Naciri J, Sashidar R, Keller P, Ratna BR (1995) Mol Cryst and Liq Cryst 263: 223
44. Clark MG, Harrison KS, Raynes EP Physics technology, vol 11. Institute of Physics Publishing, p 108
45. Lelidis I, Durand G (1993) Phys Rev Lett 48: 3822
46. Maier W, Maier G (1961) Z Naturforsch Teil A16: 262
47. Wu ST, Hsu CS (1997) SPIE 3015: 8
48. Rapini A, Popoular M (1969) J Phys (France) Colloq 30: C4
49. Tjipto-Margo B, Sullivan DE (1981)J Chem Phys 74: 3316
50. Atkins PW, Friedman RS (1997) Molecular quantum mechanics, 3rd edn. Oxford University Press, Oxford
51. Kohn W, Sham LJ (1965) Phys Rev 140: A1133
52. Jones RO, Gunnarsson O (1989) Rev Mod Phys 61: 689
53. Hohenberg P, Kohn W (1965) Phys Rev 136: B864
54. Payne M, Teter C, Joannopoulos JD (1992) Rev Mod Phys 64: 1045
55. Lin JS, Qteish A, Payne MC, Heine V (1993) Phys Rev B 47: 4174
56. Klienman L, Bylander DM (1982) Phys Rev Lett 48: 1425
57. Cepperley DM, Alder BJ (1980) Phys Rev Lett 45: 566
58. Crain J, Ackland GJ, Clark SJ (1995) Rep Prog Phys 58: 705
59. Adam CJ, Clark SJ, Ackland GJ, Crain J (1997) Phys Rev E 55: 5641
60. Osipov MA, Sluckin TJ, Cox SJ (1997) Phys Rev E 55: 464
61. Tkachenko AV (1998) Physical A 249: 380
62. Vanderschoot P (1996) J Phys 6: 1557
63. Pelcovits RA (1996) Liq Cryst 21: 361
64. Wilson MR, Allen MP (1993) Mol Phys 80: 277
65. Cross CW, Fung BM (1994) J Chem Phys 101: 6839
66. Komolkin AV, Laaksonen A, Maliniak A (1994) J Chem Phys 101: 4103
67. Feynman RP (1939) Phys Rev 56: 340
68. Pulay (1979) Mol Phys 17: 197
69. Clark SJ, Adam CJ, Ackland GJ, White J, Crain J (1997) Liq Cryst 22: 469
70. Bastiansen O, Samdal SJ (1985) Mol Struct. 128: 115
71. Häfelinger G, Regalmann C (1985) Comput J Chem 6: 368
72. Häfelinger G, Regalmann C (1987) Comput J Chem 8: 1057
73. Tsuzuki S, Tanabe K (1981) J Phys Chem 95

74. Moller C, Plesset MS (1934) Phys Rev 46: 618
75. Clark SJ, Adam CJ, Cleaver DJ, Crain J (1997) Liq Cryst 22: 477
76. Dunmur DA, Grayson M, Roy SK (1994) Liquid Crystals 16: 95
77. Adam CJ, Clark SJ, Ackland GJ, Crain J (1998) Mol Phys (in press)
78. Pertsin AJ, Kitaigorodsky AI (1986) The atom-atom potential method. Series in Chemical Physics. Springer, Berlin Heidelberg New York
79. Hsueh H, Vass H, Pu FN, Clark SJ, Poon CK, Crain J (1997) Eur Phys Lett 38: 107
80. Perova TS (1994) Adv Chem Phys 87: 427
81. Hsueh HC, Vass H, Clark SJ, Ackland GJ, Crain J (1995) Phys Rev B 51: 16750
82. Dunmur DA, Grayson M, Pickup BT, Wilson MR (1997) Mol Phys 90: 179
83. Goodman, A, Ozkabak G, Thaker N (1991) J Phys Chem 95: 9044
84. Guo H, Karplus M (1988) J Chem Phys 89: 4235
85. Handy NC, Maslen PE, Amos RD, Andrews JS, Murray CW, Laming GJ (1992) Chem Phys Lett 197: 506
86. Brenner LJ, Senekowtsch J, Wyatt RE (1993) Chem Phys Lett 215: 63
87. Albertozzi E, Zerbetto F (1992) Chem Phys 164: 91
88. Bérces A, Ziegler T (1995) Chem Phys 99: 11,417

Atomistic Simulations of Liquid Crystals

Mark R. Wilson

Department of Chemistry, South Road, University of Durham, Durham, DH1 3LE, UK
E-mail: mark.wilson@durham.ac.uk

This article reviews progress in the field of atomistic simulation of liquid crystal systems. The first part of the article provides an introduction to molecular force fields and the main simulation methods commonly used for liquid crystal systems: molecular mechanics, Monte Carlo and molecular dynamics. The usefulness of these three techniques is highlighted and some of the problems associated with the use of these methods for modelling liquid crystals are discussed. The main section of the article reviews some of the recent science that has arisen out of the use of these modelling techniques. The importance of the nematic mean field and its influence on molecular structure is discussed. The preferred ordering of liquid crystal molecules at surfaces is examined, along with the results from simulation studies of bilayers and bulk liquid crystal phases. The article also discusses some of the limitations of current work and points to likely developments over the next few years.

Keywords: Molecular mechanics, Monte Carlo molecular dynamics, atomistic simulation

1
Introduction

1.1
Force Field Methods

The rapid rise in computer power over the last ten years has opened up new possibilities for modelling complex chemical systems. One of the most important areas of chemical modelling has involved the use of classical force fields which represent molecules by atomistic potentials. Typically, a molecule is represented by a series of simple potential functions situated on each atom that can describe the non-bonded interaction energy between separate atomic sites. A further set of atom-based potentials can then be used to describe the intramolecular interactions within the molecule. Together, the potential functions comprise a *force field* for the molecule of interest.

In a typical force field the potential energy E can be represented as a sum of contributions from each structural element in the molecule.

$$E = \sum_{\text{bonds}} E_{\text{bond}} + \sum_{\text{angles}} E_{\text{angle}} + \sum_{\text{dihedrals}} E_{\text{torsion}} + \sum_{i=1}^{N} \sum_{j<i}^{N} E_{\text{nonbond}} + E_{\text{elec}}. \quad (1)$$

In Eq. (1), E_{bond} is the bond distortion energy (energy of a bond distorted from its natural bond length); E_{angle} is the angle distortion energy; E_{torsion} is the dihedral angle energy (torsional energy); E_{nonbond} is the nonbonded interaction energy (incorporating steric repulsion and the van der Waals attractive interactions) and E_{elec} the electrostatic interaction energy between two atomic sites i and j. The double sum in Eq. (1) is taken over all possible pairs of N atomic sites.

A separate potential function is used for each bond in the system. For example, a harmonic potential of the form

$$E_{\text{bond}} = \frac{1}{2} K_b (l - l_0)^2 \quad (2)$$

is often employed, where K_b is a force constant for the bond, l_0 is the equilibrium bond length and l is the actual bond length at which E_{bond} is determined. Individual bonds of a particular type (e.g., C(sp3)—C(sp3)), will have their own unique force constant and equilibrium bond length. Similarly the other energy terms in the force field will also have unique force constants [1, 2] which relate to individual structural features, such as a H—C(sp3)—C(sp3)—H dihedral angle.

It should be stressed that the terms in Eq. (1) are by no means unique and force fields can be written which incorporate a wide range of additional terms. These may include potentials to describe additional types of interaction (such as hydrogen bonding), or further terms to describe the coupling between structural elements within a molecule (such as the coupling between bond stretching and bond bending distortions). In each force field that is developed, the parameters used to describe molecular structure need to be carefully optimised to provide the best fit between experimental structural and energetic data for a particular class of molecules.

1.2
Types of Force Field

Force fields split naturally into two main classes: all-atom force fields and united atom force fields. In the former, each atom in the system is represented explicitly by potential functions. In the latter, hydrogens attached to heavy atoms (such as carbon) are removed. In their place single united (or extended) atom potentials are used. In this type of force field a CH_2 group would appear as a single spherical atom. United atom sites have the advantage of greatly reducing the number of interaction sites in the molecule, but in certain cases can seriously limit the accuracy of the force field. United atom force fields are most usually required for the most computationally expensive tasks, such as the simulation of bulk liquid crystal phases via molecular dynamics or Monte Carlo methods (see Sect. 5.1).

1.3
Force Fields for Liquid Crystal Molecules

Currently, a wide range of force fields are available in the literature. However, there have been no force fields designed specifically to model mesogenic molecules. This is partly because liquid crystal molecules are not confined to particular classes of compounds and can occur in a wide range of systems from simple organic molecules, to polymers, to metallomesogens. However, it is also true that much of the early interest in molecular modelling was strongly influenced by the desire to model biomolecular systems. Consequently, much of the early work on force fields was centred around developing accurate parameters to model the structure of peptides and nuclei acids. Amongst such force fields are those belonging to the programs CHARMM [3, 4] and AMBER [5-8]. In recent years these force fields have been extended and now contain many parameters which are sufficient to cover a range of liquid crystal molecules. This is particularly true of the all-atom AMBER94 force field [9].

Jorgensen et al. has developed a series of united atom intermolecular potential functions based on multiple Monte Carlo simulations of small molecules [10–23]. Careful optimisation of these functions has been possible by fitting to the thermodynamic properties of the materials studied. Combining these OPLS functions (Optimised Potentials for Liquid Simulation) with the AMBER intramolecular force field provides a powerful united-atom force field [24] which has been used in bulk simulations of liquid crystals [25–27].

Amongst all-atom force fields, the MM force fields of Allinger are also of particular use for studying isolated organic liquid crystal molecules. These force fields have been developed over many years, with an emphasis on predicting (as accurately as possible) all the structural and energetic data for simple organic molecules. The earlier MM1 [28] and MM2 [29] force fields have now been largely superseded by the newer MM3 [30] and MM4 [31–35] force fields. These force fields are now extremely accurate in predicting molecular structures in the gas phase and are of particular use for molecular mechanics (see Sect. 2.1). However, the drawback of these force fields is that they are extremely complicated in terms of the number of functions that are used for individual molecules. They also use the Hill potential for non-bonded van der Waals interactions [1, 30]

$$E_{\text{nonbonded}} = \varepsilon \left[-c_1 \left(\frac{r^*}{r} \right)^6 + c_2 \exp(-c_3 r / r^*) \right] \tag{3}$$

where c_1, c_2, c_3 are universal constants, r is the interatomic separation, ε is the specific energy parameter for the interaction, and r^* is the sum of van der Waals radii of the two interacting atoms. The exponential function used to evaluate the repulsive part of the non-bonded interaction between two atoms is rather expensive in terms of computer time. Consequently, the MM force fields are less suitable for computationally demanding large scale molecular dynamics or Monte Carlo simulations which involve a large number of liquid crystal molecules. For these simulations, it is usual to employ a simple Lennard-Jones 12:6 potential of the form shown in Fig. 1: trading some accuracy for considerable savings in computer time.

It is worth noting that much of the development work for the MM force fields has centred on low energy structures of molecules. Consequently, some of the force constants are less applicable to higher energy molecular structures that can occur in molecular dynamics simulations of liquid crystals.

Three other all-atom force fields have also received much recent attention in the literature MMFF94 [36–40], AMBER94 [9] and OPLS-AA [41, 42] and are becoming widely used. The latter two force fields both use non-bonded parameters which have been adjusted in order to reproduce experimental liquid phase densities and heats of vaporisation of small organic molecules. For example, OPLS-AA includes calculations on alkanes, alkenes, alcohols,

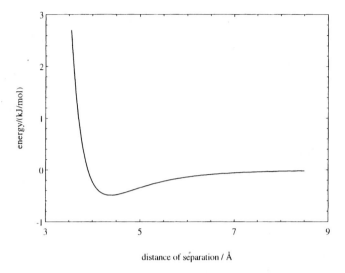

distance of separation / Å

Fig. 1. The Lennard Jones 12:6 pair potential plotted for a pair of CH_2 united atoms using the OPLS united force field. $E_{nonbond} = 4\varepsilon((\sigma/r)^{12}-(\sigma/r)^6)$, where ε is the well depth for the potential and σ is the distance at which the repulsive energy exactly cancels the attractive energy

ethers, acetals, thiols, sulfides, disulfides, aldehydes, ketones and amides in the parameterisation. Since the non-bonded parameters are the most critical ones in terms of influencing transition temperatures and densities, AMBER94 and OPLS-AA are likely to become the most important force fields for future simulations of bulk liquid crystal phases.

Finally, there are groups of liquid crystals where, at the current time, force fields are not particularly useful. These include most metal-containing liquid crystals. Some attempts have been made to generalise traditional force fields to allow them to cover more of the periodic table [40, 43]. However, many of these attempts are simple extensions of the force fields used for simple organic systems, and do not attempt to take into account the additional strong polarisation effects that occur in many metal-containing liquid crystals, and which strongly influence both molecular structure and intermolecular interactions.

2
Simulation Methods for Liquid Crystal Molecules

2.1
Molecular Mechanics

Molecular mechanics is the simplest and the most commonly used molecular modelling technique. It focuses on the low energy structures of molecules which correspond to energy minima in the classical force field used to represent the molecule [44]. A typical molecular mechanics study consists of the following steps [45]:

1. build a trial molecular geometry;
2. minimise the energy of the trial geometry to provide an energy minimised structure;
3. use a search algorithm to undertake a search for other energy minima by adjusting the molecular structure (e.g. by carrying out rotations about dihedral angles and carrying out energy minimisations for each new trial geometry).

The end result of this process is a series of energy minima for the molecule of interest. An appropriate weight can be assigned to each conformation using the usual Boltzmann weighting

$$P_j = \frac{\exp(-\Delta E_j/k_B T)}{\sum\limits_{i=1}^{N_{conf}} \exp(-\Delta E_i/k_B T)} \tag{4}$$

where T is the temperature and ΔE_i is the energy of conformation i relative to the lowest energy conformation found. For many large liquid crystal molecules the number of accessible structures can be very large indeed. In these cases, specialist techniques for finding conformers are required. However, the problem of guaranteeing that all important conformations have been found is a difficult one, and no one technique can yet guarantee to find all the minima on a multi-dimensional energy surface.

Several excellent reviews of molecular mechanics exist in the literature and the reader is directed to these for further details of the methods commonly in use [1, 2, 46–47].

2.2
Molecular Dynamics

One of the ways of avoiding the multiple-minima problem that occurs with molecular mechanics is by turning to statistical mechanical techniques which provide ways of sampling the force field energy E without restriction to individual energy minima. The two techniques that achieve this for molecular systems are molecular dynamics (MD) and Monte Carlo simulation (MC). MD techniques for molecular systems have been widely reviewed [48–50]. The MD method relies on solving Newton's equations of motion for molecules interacting via an appropriate potential function of the form given in Eq. (1). At the start of the simulation atoms will be given a velocity sampled from a Maxwell-Boltzmann distribution; they will also experience a force given by

$$F_i = -\nabla E_i. \tag{5}$$

By breaking time down into small intervals δt, the equations of motion can then be solved directly using finite difference algorithms [48]. In the simplest form of MD the total energy of the molecular system is a conserved quantity. However, it is equally possible to carry out MD at constant temperature by employing one of a number of available thermostat algorithms [51]. When simulating bulk materials periodic boundary conditions are employed to remove any edge effects, together with an appropriate statistical mechanical ensemble. For liquid crystals the most common ensembles used are the constant NVE (microcanonical), constant NVT (canonical) and constant NpT (isobaric-isothermal) ensemble. The latter is of particular use in simulating smectic liquid crystals because the size and shape of the periodic box enclosing the fluid is allowed to vary to avoid smectic layers forming which are incommensurate with the boundary conditions.

2.3
Molecular Monte Carlo

Monte Carlo simulations derive their name from the random numbers which are used in this form of calculation. In molecular Monte Carlo, the random numbers are used to generate appropriate energy states E_i, usually via the Metropolis algorithm [52–54]. A random change is made to the molecular coordinates and the energy of the new configuration E_{new} is calculated and compared to the energy of the old configuration E_{old}. Random changes to Cartesian coordinates can be made, but usually it is most efficient to make changes to the internal coordinates of a molecule [55]. If $E_{new} < E_{old}$, then the trial move is accepted. If $E_{new} > E_{old}$, the move is accepted if $\exp(-(E_{new} - E_{old})/k_BT)$ is greater than a random number between 0 and 1. Consequently, over the course of a MC simulation, moves are accepted with a probability of $\exp(-(E_{new} - E_{old})/k_BT)$. Monte Carlo methods tend to sample phase space much more efficiently for single molecules than molecular dynamics [55, 56]. They should therefore be the method of choice for single molecule calculations on liquid crystal systems. However, it is harder to produce a generalised method for carrying out molecular Monte Carlo than it is for molecular dynamics. So many standard molecular modelling packages only provide molecular mechanics and molecular dynamics and not Monte Carlo. Recently, several combined molecular dynamics and Monte Carlo methods have appeared in the literature [56]. These are likely to be of particular use for liquid crystal materials, which often contain many internal degrees of freedom and require large equilibration times.

In contrast to the single molecule case, Monte Carlo methods tend to be rather less efficient than molecular dynamics in sampling phase space for a bulk fluid. Consequently, most of the bulk simulations of liquid crystals described in Sect. 5.1 use molecular dynamics simulation methods.

3
Simulation of Isolated Liquid Crystal Molecules

3.1
Structure Determination Using Molecular Mechanics

The determination of molecular structure is the simplest application of molecular modelling. Most liquid crystal molecules are found to have a number of conformations which are thermally accessible at room temperature.

For example mesogens of the homologous series 4-(*trans*-4-*n*-alkylcyclohex-yl)benzonitrile (PCH series in Scheme 1) possess energy minima correspond-ing to *trans* and *gauche* dihedral angles in the chain and rotation about the bond linking the cyclohexyl and phenyl rings.

The gas phase structures of PCH3 and PCH5 have been determined [45] using the MM3 force field [30] and the Macromodel molecular modelling package [57]. In its minimum energy conformation the phenyl ring lies in the symmetry plane of the cyclohexyl ring, the two possible torsional angles corresponding to label **a** in Scheme 1 can take values of −118° and +118°. However, the energy barrier to rotation about **a** is only 7.5 kJ mol^{-1}. So at room temperature (at 298 K, RT ≈ 2.5 kJ mol^{-1}) we expect reasonable conformational flexibility about this dihedral, but not completely free rotation. Dihedral **b** represents an important structural feature that occurs in a number of liquid crystal molecules. The torsional angle potential shown in Fig. 2 illustrates that the *all-trans* chain prefers not to lie in the symmetry plane of the cyclohexyl ring (as in Scheme 2, structure 3), but has two possible conformations of equal energy (Scheme 2, structures 1 and

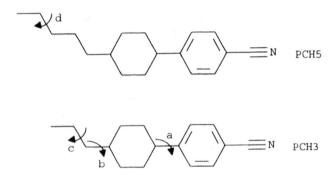

Scheme 1. Structures of 4-(*trans*-4-*n*-pentylcyclohexyl)benzonitrile (PCH5) and 4-(*trans*-4-*n*-prop-ylcyclohexyl)benzonitrile (PCH3)

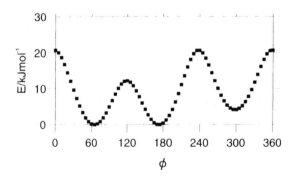

Fig. 2. The torsional angle potential for dihedral angle **b** of PCH3. Data supplied by Dr M R Wilson, University of Durham from MM3 calculations [45]

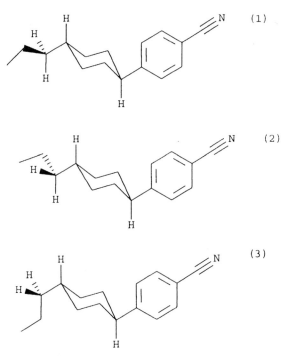

Scheme 2. Lowest energy conformations for dihedral **b** of PCH3. Structures (**1**), (**2**) and (**3**) in turn correspond to the three energy minima in Fig. 2

2). These two (equivalent) conformations are also found experimentally in the crystal structures of liquid crystals with this molecular fragment [58–60].

Figure 3 plots the torsional angle potential for dihedral **d** in PCH5. It shows behaviour which is generic for alkyl chain dihedrals. The *gauche* conformers

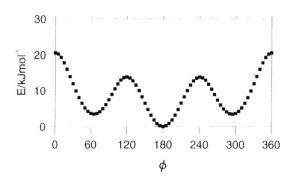

Fig. 3. The torsional angle potential for dihedral angle **d** of PCH5. Data supplied by Dr M R Wilson, University of Durham from MM3 calculations [45]

lie approximately 3.4 kJ mol^{-1} in energy above the *trans* conformer and the barriers to rotation are respectively 13.9 kJ mol^{-1} (for two barriers) and 20.5 kJ mol^{-1}. The *gauche/trans* energy gap is therefore quite small, and we expect significant populations of *gauche* conformations at room temperature.

For a given conformer (**1**) or (**2**) in Scheme 2, the dihedral angle c of PCH3 in Scheme 1 gives rise to two *gauche* conformations which are unequal in energy. In this case one conformation leads to the chain colliding with the cyclohexane ring [45].

3.2
The Role of the Nematic Mean Field

Molecular mechanics force fields have largely been parameterised using the best available data from the gas phase and (in some cases) from liquid phase or solution data. The question therefore arises as to how applicable molecular mechanics force fields are to predicting structures of molecules in the liquid crystal phase. There is now good evidence from NMR measurements that the structure of liquid crystal molecules change depending on the nature of their

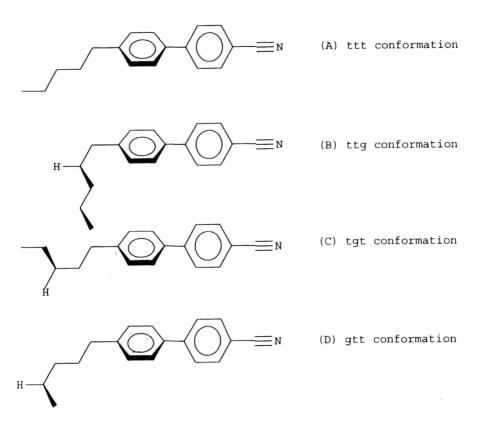

Scheme 3. Lowest energy conformations for the mesogen 5CB (4-*n*-pentyl-4′-cyanobiphenyl)

environment [61]. This can be seen clearly through molecular field theory calculations which have attempted to correlate predicted bond order parameters with those derived from deuterium couplings [62] for the molecule 4-*n*-pentyl-4′-cyanobiphenyl (5CB). The results clearly suggest that individual conformations in which the molecule remains elongated are preferentially selected by the nematic field. In Scheme 3, this corresponds to structures A (*all-trans* conformation) and structure C (*tgt* conformation of dihedrals in the alkyl chain) being preferred. In principle a correction factor could be included in all molecular mechanics calculations to take the effects of (say) a nematic mean field into account, this would correspond to an additional mean field term in Eq. (1) which would be dependent on molecular conformation. This term would then be included during the energy minimisation steps. As yet this scheme has not been implemented in any molecular mechanics packages; it has however been used in Monte Carlo calculations for single molecules (see Sect. 3.3) [63].

3.3
Monte Carlo Simulations of Isolated Liquid Crystal Molecules

Monte Carlo simulations can be carried out particularly efficiently for isolated molecules. If internal coordinates are implemented by means of a Z-matrix, a general MC sampling scheme can be produced which involves making random changes to the bond lengths, bond angles and proper dihedral angles defined in the Z-matrix [63]. This method has been used for 5CB in the gas phase and also in the presence of a conformationally dependent potential of mean torque U_{ext}, that is used to mimic the influence of the nematic mean field on molecular conformation. The results of this technique are very promising. As the strength of the potential of mean torque is increased, the simulations see the preferential selection of the most elongated conformers (as seen in molecular field theory in Sect. 3.2): the preferred *gauche* conformer changes from the *gtt* conformer (**B** in Scheme 3) to the tgt conformer (**C** in Scheme 3). In reference [63], the simplest form of U_{ext} is used

$$U_{ext} = -v\xi(\Gamma)P_2(\cos\beta) \qquad (6)$$

where for a conformation Γ, β is the angle between the molecular long axis and the director, v indicates the strength of the nematic mean field and $\xi = a(\Gamma) - (b(\Gamma)+c(\Gamma))/2$, where a, b, c are the semi-axis lengths of the molecular inertia spheroid. However, it is also possible to use more sophisticated forms for U_{ext}, which will take account of molecular biaxiality. One form for U_{ext} which seems to be of general applicability takes into account the surface of the liquid crystal molecule [64]

$$U_{ext} = v\int_S P_2\cos\Psi_{ns}\,ds. \qquad (7)$$

Here, Ψ_{ns} is the angle between the director and a vector normal to the molecular surface, and the integral is taken over all points on the surface S. There are a number of possible ways of defining S. In quantum mechanical calculations S is naturally defined in terms of an appropriate contour of the electron density. However, the simplest way of viewing the molecular surface is by using van der Waals spheres on each atom [64]. Although moderately expensive in terms of computer time, this method can naturally be included in the force field energy used in single molecule Monte Carlo simulations.

4
Interaction of Liquid Crystal Molecules with a Surface Potential

4.1
The Need to Study Liquid Crystal Molecules at Surfaces

Good alignment of liquid crystal molecules at the liquid crystal/substrate interface is of major importance in the fabrication of display devices. However, the actual mechanism of alignment is quite poorly understood. In recent years it has become possible to image liquid crystals on surfaces using high resolution scanning tunnelling microscopy (STM) [65–67]. However, "clean" surfaces of graphite [65, 66] and MoS_2 [67] are somewhat removed from the typical "rubbed" polymer surfaces which are used as alignment layers in liquid crystal displays. There is currently a debate regarding the influence of rubbing on the polymer alignment layer. Rubbing is known to produce grooves in the polymer [68, 69] (which may encourage alignment of liquid crystal molecules), but rubbing may also induce strong alignment of the orientation of the polymer chains [70]. This would in turn produce alignment of an adsorbed liquid crystal. There is therefore considerable interest in using computer simulations to study the mechanism of surface alignment for LC systems.

4.2
Simulations of Mesogens Adsorbed on Surfaces

Yoneya and Iwakabe have presented preliminary simulations [71] of a monolayer of eight 4-n-octyl-4'-cyanobiphenyl (8CB) at 300 K on the basal plane of graphite using the GROMOS force field [72]. For short runs of 45 ps they conclude that the alkyl chain provides the main contribution to the adsorption energy. The same authors also consider 8CB on a polyimide monolayer [73]. The actual system studied consisted of a three-layer sandwich composed of liquid crystal/polyimide oligomer/graphite. Over the course of 500 ps simulations the authors are able to observe 8CB molecules aligning parallel to the polyimide chain direction with an almost zero pretilt.

Cleaver and co-authors [74, 75] have also looked at 8CB on graphite carrying out both detailed energy minimisation work and molecular dynamics simulations using periodic boundary conditions in two directions. (Zero kelvins) energy minimisations of 8CB lead to a most favourable energy of adsorption for an eight-molecule unit-cell [75] where molecules pack in a head-to-tail arrangement within strips, a result which is wholly consistent with STM studies. Molecular dynamics simulations of systems of 72 and 288 molecules in an 8CB monolayer suggest that molecules remain ordered in strips forming a two-dimensional smectic phase at 150 K but lose the details of the eight-molecule unit-cell. The presence of a model potential above the monolayer, designed to model the influence of a STM tip, has the effect of confining the monolayer and keeping the eight-molecule unit cell structure. A similar effect also occurs when a second 8CB layer is placed on top of the original monolayer, though the effect is less marked in this case. It is also interesting to note that, in common with other similar studies, the Lennard-Jones interactions are found to completely dominate the electrostatic contribution to the total interaction between the adsorbate and the surface. The electrostatic interactions contribute less than 2% to the total surface energy. However, electrostatics are likely to be more important when considering bulk liquid crystals in contact with a surface. The longer range nature of the electrostatic interactions will make them relatively more significant in this case.

Binger and Hanna [76] have considered the alignment of 4-n-octyl-2-flurophenyl-4-n-octyloxybiphenyl-4'-carboxylate (MBF) and 8CB at room temperature on the (110) surface of polyethylene and the (100) surface of nylon. They find that alignment of both molecules is dominated by the behaviour of the flexible tails that prefer to align parallel between two polymer chains. This forces the mesogenic core to straddle one or more chains. This is illustrated in Fig. 4 for the specific case of MBF on the (110) surface of

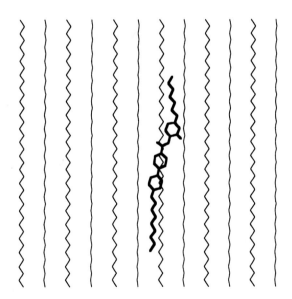

Fig. 4. An example configuration of a MBF molecule on the 110 crystal face of polyethylene taken from the simulation data of [76]. Data supplied by Dr D R Binger and Dr. S Hanna, University of Bristol

polyethylene, where both MBF chains are shown lying between a pair of polymer chains and the core of the molecule straddles a single polymer chain.

With clusters of eight MBF molecules, the orientational freedom of individual molecules becomes restricted, and, as a consequence of this, the distribution of core orientations changes to favour strongly orientations where the molecular core straddles only one polymer chain.

When suitably doped, MBF can form a surface-stabilised-ferroelectric smectic-C* (SSFLC) structure. Using simple assumptions regarding core orientations, Binger and Hanna are able to place an upper limit on the SSFLC cone angle for MBF of 30°.

5
Simulation of Bulk Liquid Crystal Phases

5.1
Phase Behaviour of Thermotropic Liquid Crystals

In recent years much progress has been made in the simulation of bulk liquid crystal phases using simple potential models. Both hard [77–82] (e.g. hard ellipsoids, hard spherocylinders and chains of hard spheres) and soft anisotropic models (e.g. the Gay-Berne potential [83–86]) have been used successfully to simulate a range of different phases. These simulations have also allowed the calculation of bulk properties such as elastic constants [87, 88] and transport properties [89], which have great relevance for the use of thermotropic liquid crystal materials in displays applications. The question therefore arises as to whether or not atomistic models can also be used to simulate bulk phases in the same way as these simpler models. The drawback of atom-based models is that they are extremely expensive in terms of computer time. As a general guide, the amount of computer time needed increases with the square of the total number of atomic sites in the system. (This can sometimes be improved by truncating short range potentials and using neighbour lists to reduce the number of pairwise interactions which must be computed). Despite this drawback, a number of atomistic simulations of bulk phases have been carried out for relatively small system sizes and fairly short simulation times. The results of these studies point to this being an extremely fruitful area of research, but the conclusions drawn below must all be considered as preliminary until longer simulations can be carried out with larger system sizes.

Table 1 provides a summary of the main atomistic simulations to date [25–27, 90–104]. Simulation runs of around 400 ps seem quite sufficient to equilibrate the internal molecular structure of most mesogens provided that the barriers to internal rotation about bonds are not too high (< 20 kJ mol^{-1}). However, the growth and decay of the orientational order parameter for an aligned system can take place over much longer timescales. Interpolation from the results of detailed studies of Gay-Berne mesogens suggest that run lengths of 5–10 ns will be required to see the growth of a nematic phase from the isotropic liquid and runs of around 0.5–1.0 ns will be required to move from

Table 1. Atomistic simulations of bulk liquid crystals

Acronym	Molecule	Ref.	N_m	Study
EBBA	4-ethoxybenzylidene-4′-n-butylaniline	[90]	60	NpT MC (all atom)
5CB	4-n-pentyl-4′-cyanobiphenyl	[91]	64	NpT MD
		[98]	80	NpT MD
		[100]	75	NVE MD
pHB	phenyl-4-(4-benzoyloxy-)benzoyloxy benzoate	[92]	8,16	NVT MD (all atom)
CCH5	4-(trans-4-n-pentylcyclohexyl)cyclohexylcarbonitrile	[25–27]	128	NpT MD
THE5	hexakis(pentyloxy)triphenylene	[93]	54	NVT MD
nOCB	4-alkoxy-4′cyanobiphenyl	[94, 95]	64	NpT MD
(n=5–8)			125,	NpT MD
5OCB		[103]	144	
PCH5	4-(trans-4-n-pentylcyclohexyl)benzonitrile	[96, 101, 102]	50, 100	NpT MD
HBA	tetramer of 4-hydroxybenzoic acid	[97]	125	NpT MD (all atom)
2MBCB	(+)-4-(2-methylbutyl)-4′-cyanobiphenyl	[99]	32	NpT MD
5,5-BBCO	4,4′-di-n-pentyl-bibicyclo[2.2.2]octane	[104]	64, 125	NVT MD

one state point to another in a nematic phase. Most early atomistic studies fell well short of these run lengths and had to rely on studying the results from short simulations in which the director remained approximately constant. Recently however, McBride et al. have demonstrated the thermodynamic stability of an atomistic nematic phase by growing it directly from a completely isotropic liquid [104]. For the molecule 5,5-BBCO, two independent quenches from an isotropic liquid were carried out with the growth of the nematic phase taking place over 6 ns in one instance and 10 ns in the other. The difference in timescale seems to be dictated mainly by the degree to which the system is quenched below the phase transition.

The orientational order parameter for a liquid crystal can be measured by first calculating the ordering tensor

$$\mathbf{Q}_{\alpha\beta} = \frac{1}{N_m} \sum_{j=1}^{j\ N_m} \frac{3}{2} u_{j\alpha} u_{j\beta} - \frac{1}{2} \delta_{\alpha\beta}, \quad \alpha, \beta = xyz \tag{8}$$

for a configuration of N_m molecules, where $\mathbf{u}_j$ is the vector describing the orientation of the part of the molecule of interest. Diagonalisation of $\mathbf{Q}$ yields the uniaxial order parameter S_2 as the largest eigenvalue (λ_+) of Q and the director as the corresponding eigenvector. Alternatively, S_2 can be obtained as $-2 \times$ the middle eigenvalue of $\mathbf{Q}$. This definition of S_2 gives very similar values to λ_+ in a uniaxial phase, but has the advantage of fluctuating either side of zero in the isotropic phase. This avoids the problems with finite system size that always give rise to small positive values of λ_+ in an isotropic liquid. Separate values of S_2 can easily be measured for individual parts of a molecule, such as the bonds in an alkyl chain or the long axis of a rigid molecular segment. However, a convenient "all-molecule" definition for S_2 arises from defining $\mathbf{u}_j$ as the molecular long axis provided by the diagonalisation of the inertia tensor [26]

$$\mathbf{I}_{\alpha\beta} = \sum_i m_i \left(r_i^2 \delta_{\alpha\beta} - r_{i\alpha} r_{i\beta} \right), \quad \alpha, \beta = xyz \tag{9}$$

where atomic masses are given by m_i, and each atomic distance vector $\mathbf{r}_i$ is measured relative to the molecular centre of mass. Figure 5 shows an example of the growth of S_2 measured from the inertia tensor-defined for the long axis of 5,5-BBCO [104]. Snapshots illustrating the order of molecules in the istropic liquid and the mesophase formed at lower temperatures are shown in Fig. 6.

Wilson [27] has examined the use of the polarisability tensor to define the long molecular axis for the molecule trans-4-(trans-4-n-pentylcyclohexyl)cyclohexylcarbonitrile (CCH5). For CCH5, the direction of the long axis defined by polarisability tensor and the inertia tensor are quite different. However, the values of S_2 and the form of the orientational distribution function are almost identical for the two definitions of the long molecular axis. It is also interesting to note that the order parameter calculated for the cyano group in CCH5 (or any other sub-unit within the molecule) is rather less than

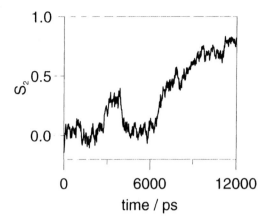

Fig. 5. The growth of orientational order for the 64 molecules of 5,5-BBCO during a 300 K quench starting from the isotropic liquid. Data from [104], supplied by C McBride, University of Durham

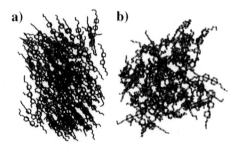

Fig. 6a,b. Snapshots of 5,5-BBCO molecules in model: **a** nematic; **b** isotropic liquid phases. Data from [104], supplied by C McBride, University of Durham

that obtained from either of the two whole-molecule definitions of S_2. This indicates that considerable caution should be exercised in using experimental order parameters that are obtained by optical methods that strongly rely on particular chromophores within a molecule.

The question arises as to how useful atomistic models may be in predicting the phase behaviour of real liquid crystal molecules. There is some evidence that atomistic models may be quite promising in this respect. For instance, in constant pressure simulations of CCH5 [25, 26] stable nematic and isotropic phases are seen at the right temperatures, even though the simulations of up to 700 ps are too short to observe spontaneous formation of the nematic phase from the isotropic liquid. However, at the present time one must conclude that atomistic models can only be expected to provide qualitative data about individual systems rather than quantitative predictions of phase transition temperatures. Such predictions must await simulations on larger systems, where the system size dependency has been eliminated, and where constant pressure MD can be carried out with the next generation of all-atom force fields with intermolecular potentials carefully tuned to provide accurate liquid state densities of small molecules (see Sect. 1.3).

5.2
Molecular Structure in Thermotropic Liquid Crystal Phases

One of the key advantages of atomistic models over simpler single site potentials is that they allow the study of molecular structure in the mesophase. A number of workers have looked at this in some detail [25, 26, 98, 104] through dihedral angle distributions, the relative proportions of molecular conformers and effective torsional angle potentials. The dihedral angle distributions $S(\phi)$ that can be calculated directly in liquid and liquid crystal phases are linked to the effective torsional angle $E_{eff}(\phi)$ (conformational free energy) by the relation

$$S(\phi) = C \exp\left(-\left(\frac{E_{eff}(\phi)}{k_B T}\right)\right) \tag{10}$$

where C is a normalization factor. $E_{eff}(\phi)$ can be split into contributions from the internal molecular structure, $E_{int}(\phi)$, and "external" contributions from the surrounding molecules $E_{ext}(\phi)$. $E_{ext}(\phi)$ is strongly influenced by changes in molecular environment which occur when the orientational order of surrounding molecules change (e.g. at the isotropic/nematic phase transition). Mesogens have been found to change their average molecular shape and become "longer" and "thinner" as they move from an isotropic liquid into a nematic phase. Individual conformers are chosen in preference to others. For the molecule CCH5 [26] the calculated order of chain conformers is ttt > tgt > ttg > gtt, in agreement with molecular field theory predictions for the similar molecule 5CB (see Sects. 3.2 and 3.3). However, the specific influence of the structural mean field surrounding a molecule depends strongly on the details of molecular shape. So for CCH5 (with one C_5 alkyl chain) the mean field raises the *gauche-trans* energy difference for the two odd dihedral angles in the chain, but for 5,5-BBCO (with two C_5 alkyl chains) it is the even numbered dihedrals that experience a reduction in the *gauche-trans* energy difference.

5.3
Calculation of Bulk Properties for Thermotropic Phases

A number of bulk simulations have attempted to study the dynamic properties of liquid crystal phases. The simplest property to calculate is the translational diffusion coefficient D, that can be found through the Einstein relation, which applies at long times t:

$$D = \frac{1}{2t}\left\langle |\mathbf{r}_i(t) - \mathbf{r}_i(t_0)|^2 \right\rangle. \tag{11}$$

Here the vector r_i represents the centre of mass position, and D is usually averaged over several time origins t_0 to improve statistics. Values for D can be resolved parallel and perpendicular to the director to give two components $(D_{//}, D_{\perp})$, and actual values are summarised for a range of studies in Table 3 of [45]. Most studies have found diffusion coefficients in the 10^{-9} m^2 s^{-1} range with the ratio $D_{//}/D_{\perp}$ between 1.59 and 3.73 for calamitic liquid crystals. Yakovenko and co-workers have carried out a detailed study of the reorientational motion in the molecule PCH5 [101]. Their results show that conformational molecular flexibility plays an important role in the dynamics of the molecule. They also show that cage models can be used to fit the reorientational correlation functions of the molecule.

In simple single-site liquid crystal models, such as hard-ellipsoids or the Gay-Berne potential, a number of elegant techniques have been devised to calculate key bulk properties which are useful for display applications. These include elastic constants for nematic systems [87, 88]. However, these techniques are dependent on large systems and long runs, and (at the present time) limitations in computer time prevent the extension of these methods to fully atomistic models.

5.4
Simulation of Bilayers and Lyotropic Systems

Closely related to the simulation of bulk liquid crystal phases is the study of monolayer and bilayer systems. There has been much interest in this area in recent years [105–110], particularly in the study of bilayer systems where the simulations also attempt to simulate directly the presence of water. The latter act as model membranes and are of major interest in computational biochemistry. The most popular systems for study have been lipid bilayers and these have been thoroughly reviewed by Pastor [109] and Tobias et al. [110]. For these systems, where the packing of chains strongly influences the behaviour of the model, it has been found that united atom force fields are not sufficient and hydrogens must be included specifically. There is also general agreement about the need to carry out simulations at constant pressure and to use the Ewald sum to take properly into account the long range charge-charge interactions, rather than truncate them with a spherical cut-off (as usually occurs for Lennard-Jones 12:6 potentials). Current state-of-the-art calculations appear to be able to reproduce average structures of both gel and liquid crystal membrane phases quite successfully using simulations of up to 10 ns in duration. However, there are still many deficiencies with these models, particularly at the bilayer/water interface [110].

5.5
Use of Hybrid Atomistic Models

An interesting variant of fully atomistic simulations arises when rigid sections of the model mesogen are replaced by simpler potentials. Cross and Fung [111] have replaced the biphenyl core of some n-alkylcyanobiphenyls with a large sphere situated in the centre of each phenyl ring. This provides large savings in computer time over united atom models of the same molecules. Alternatively, soft anisotropic potentials of the Gay-Berne form can be used to replace rigid parts of mesogenic molecules. Wilson [112] has carried out extensive simulations of a liquid crystal dimer composed of two rigid Gay-Berne units linked by a semi-flexible alkyl chain using the GBMOL package [113]. The

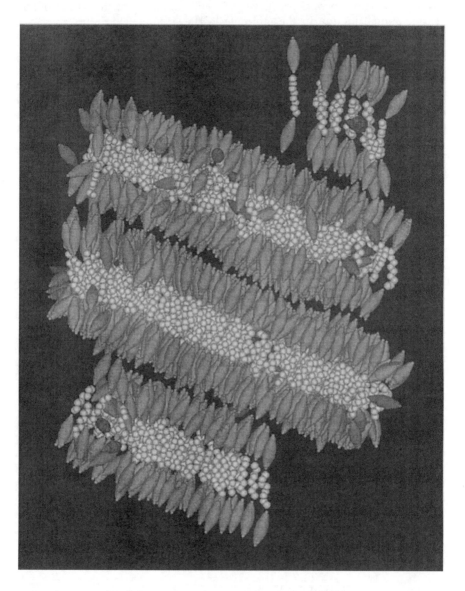

Fig. 7. Snapshots from a smectic-A phase of liquid crystal dimer molecules composed of two Gay-Berne mesogens linked by a flexible alkyl chain. Data from [112] supplied by Dr M R Wilson, University of Durham

savings afforded by the Gay-Berne potentials allowed simulations of 512 molecules for times of up to 6.4 ns. Using this model it proved possible to grow a smectic-A phase (Fig. 7) directly from an isotropic liquid and pin the phase transition down to approximately 30 K. This hybrid Gay-Berne/ Lennard-Jones approach has also been used to carry out the first atom-based simulations of a main-chain liquid crystal polymer in which the mesophase has been grown spontaneously from the liquid [114]. The large odd-even effects, seen experimentally for many polymer systems as alkyl chain lengths vary, are convincingly demonstrated in this study.

6
Perspectives

The rapid rise in computer speed over recent years has led to atom-based simulations of liquid crystals becoming an important new area of research. Molecular mechanics and Monte Carlo studies of isolated liquid crystal molecules are now routine. However, care must be taken to model properly the influence of a nematic mean field if information about molecular structure in a mesophase is required. The current state-of-the-art consists of studies of (in the order of) 100 molecules in the bulk, in contact with a surface, or in a bilayer in contact with a solvent. Current simulation times can extend to around 10 ns and are sufficient to observe the growth of mesophases from an isotropic liquid. The results from a number of studies look very promising, and a wealth of structural and dynamic data now exists for bulk phases, monolayers and bilayers. Continued development of force fields for liquid crystals will be particularly important in the next few years, and particular emphasis must be placed on the development of all-atom force fields that are able to reproduce liquid phase densities for small molecules. Without these it will be difficult to obtain accurate phase transition temperatures. It will also be necessary to extend atomistic models to several thousand molecules to remove major system size effects which are present in all current work. This will be greatly facilitated by modern parallel simulation methods that allow molecular dynamics simulations to be carried out in parallel on multi-processor systems [115].

As computer power continues to increase over the next few years, there can be real hope that atomistic simulations will have major uses in the prediction of phases, phase transition temperatures, and key material properties such as diffusion coefficients, elastic constants, viscosities and the details of surface adsorption.

7
References

1. Allinger NL (1976) Rev Phys Org Chem 13: 1
2. Burkert O, Allinger NL (1982) Molecular mechanics. ACS monograph 177. American Chemical Society, Washington, DC

3. Gelin BR, Karplus M (1979) Biochemistry 18: 1256
4. Brooks BR, Bruccoleri RE, Olafson BD, States DJ, Swaminathan S, Karplus M (1983) J Comput Chem 4: 187
5. Singh UC, Weiner PK, Caldwell J, Kollman PA (1987) AMBER 3.0. University of California, San Francisco
6. Weiner PK, Kollmann PA (1981) J Comput Chem 2: 287
7. Weiner SJ, Kollman PA, Nguyen DT, Case DA (1986) J Comput Chem 7: 230
8. Weiner SJ, Kollman PA, Case DA, Singh UC, Ghio C, Alagona G, Profeta S, Weiner P (1984) J Am Chem Soc 106: 765
9. Cornell WD, Cieplak P, Bayly CI, Gould IR, Merz KM Jr, Ferguson DM, Spellmeyer DC, Fox T, Caldwell JW, Kollman PA (1995) J Am Chem Soc 117: 5179
10. Kaminski G, Duffy EM, Matsui T, Jorgensen WL (1994) J Phys Chem 98: 13077
11. Duffy EM, Jorgensen WL (1994) J Am Chem Soc 116: 6337
12. Jorgensen WL, Laird ER, Nguyen TB, Tiradorives J (1993) J Comput Chem 14: 206
13. Jorgensen WL, Nguyen TB (1993) J Comput Chem 14: 195
14. Briggs JM, Matsui T, Jorgensen WL (1990) J Comput Chem 11: 958
15. Jorgensen WL, Briggs JM (1988) Molec Phys, 63: 547
16. Jorgensen WL, Tiradorives J (1988) J Am Chem Soc 110: 1666
17. Jorgensen WL (1986) J Phys Chem 90: 6379
18. Jorgensen WL, Gao J (1986) J Phys Chem 90: 2174
19. Jorgensen WL (1986) J Phys Chem 90: 1276
20. Jorgensen WL, Madura JD (1985) Molec Phys 56: 1381
21. Jorgensen WL, Swenson CJ (1985) J Am Chem Soc 107: 1489
22. Jorgensen WL, Swenson CJ (1985) J Am Chem Soc 107: 569
23. Jorgensen WL, Madura JD, Swenson CJ (1984) J Am Chem Soc 106: 6638
24. Jorgensen WL, Tirado-Rives J (1988) J Am Chem Soc 110: 1666
25. Wilson MR, Allen MP (1991) Molec Cryst Liq Cryst 198: 465
26. Wilson MR, Allen MP (1992) Liq Cryst 12: 157
27. Wilson MR (1996) J Molec Liq 68: 23
28. Allinger NL, Tribble MT, Miller MA, Wertz DH (1971) J Am Chem Soc 93: 1637
29. Allinger NL (1977) J Am Chem Soc 99: 8127
30. Allinger NL, Yuh YH, Lii J (1989) J Am Chem Soc 111: 8551
31 Allinger NL, Chen KS, Lii JH (1996) J Comput Chem 17: 642
32 Nevins N, Chen KS, Allinger NL (1996) J Comput Chem 17: 669
33 Nevins N, Lii JH, Allinger NL (1996) J Comput Chem 17: 695
34 Nevins N, Allinger NL (1996) J Comput Chem 17: 730
35 Allinger NL, Chen KS, Katzenellenbogen JA, Wilson SR, Anstead GM (1996) J Comput Chem 17: 747
36 Halgren TA (1996) J Comput Chem 17: 490
37 Halgren TA (1996) J Comput Chem 17: 520
38 Halgren TA (1996) J Comput Chem 17: 553
39 Halgren TA, Nachbar RB (1996) J Comput Chem 17: 587
40 Halgren TA (1996) J Comput Chem 17: 616
41. Jorgensen WL, Maxwell DS, Tirado-Rives J (1996) J Am Chem Soc 118: 11225
42. Jorgensen WL, McDonald NA (1998) Theochem – J Molec Struct 424: 145
43. CAChe molecular moldelling program (1996) Oxford Molecular Group Inc
44. In molecular mechanics the force field energy E is usually termed the *steric energy* of the molecule
45. Wilson MR (1998) Molecular modelling. In: Demus D, Goodby JW, Gray GW, Spiess HW, Vill V (eds) Handbook of liquid crystals, vol 1. Fundamentals, chap III. Wiley-VCH, Weinheim, p 3
46. Boyd DB, Lipkowitz KB (1982) J Chem Education 59: 269
47. Wilson MR (1988) PhD Thesis, University of Sheffield
48. Allen MP, Tildesley DJ (1987) Computer simulation of liquids, chap 3. Oxford University Press, Oxford

49. Rapaport DC (1995) The art of molecular dynamics simulation. Cambridge University Press, Cambridge
50. Frenkel D (1996) Understanding molecular simulation. Academic Press, San Diego
51. Hoover WG (1985) Phys Rev A31: 1695
52. Jorgensen WL (1983) J Phys Chem 87: 5304
53. Levesque D, Weis JJ, Hansen JP (1984) In: Binder K (ed) Topics in current physics 36 – applications of the Monte Carlo method in statistical physics, chap 2. Springer, Berlin Heidelberg New York
54. Allen MP, Tildesley DJ (1987) Computer simulation of liquids, chap 4. Oxford University Press, Oxford
55. Leggetter S, Tildesley DJ (1989) Molec Phys 68: 519
56. Guarnieri F, Still WC (1994) J Comput Chem 15: 1302
57. MacroModel V3.5X (1992) Interactive molecular modeling system. Department of Chemistry, Columbia University, New York
58. Haase W, Paulus H (1983) Molec Cryst Liq Cryst 100: 111
59. Haase W, Paulus H (1983) Molec Cryst Liq Cryst 92: 237
60. Haase W, Loub J, Paulus H, Mokhles MA, Ibrahim IH (1991) Molec Cryst Liq Cryst 197: 57
61. Emsley JW, Luckhurst GR, Stockley CP (1981) Molec Phys 44: 565
62. Emsley JW, Luckhurst GR, Stockley CP (1982) Proc R Soc Lond A 381: 139
63. Wilson MR (1996) Liq Cryst 21: 437
64. Ferrarini A, Moro GJ, Nordio PL, Luckhurst GR (1992) Molec Phys 77: 1
65. Foster JS, Frömmer JE (1988) Nature 333: 542
66. Smith DPE, Horber JKH, Binnig G, Nejoh H (1990) Nature 344: 641
67. Hara M, Iwakabe Y, Tochigi K, Sasabe H, Garito AF, Yamada A (1990) Nature 344: 228
68. Berreman DW (1972) Phys Rev Lett 28: 1683
69. Berreman DW (1973) Mol Cryst Liq Cryst 23: 215
70. Castellano JA (1983) Mol Cryst Liq Cryst 94: 33
71. Yoneya M, Iwakabe Y (1995) Liq Cryst 18: 45
72. van Gunsteren WF, Berendsen HJC (1987) GROMOS Manual, BIOMOS BV
73. Yoneya M, Iwakabe Y (1996) Liq Cryst 21: 347
74. Cleaver DJ, Tildesley DJ (1994) Molec Phys 81: 781
75. Cleaver DJ, Callaway MJ, Forester T, Smith W, Tildesley DJ (1995) Molec Phys 86: 613
76. Binger DR, Hanna S (1997) Mol Cryst Liq Cryst 302: 63
77. For a review of the early work in this field see: Allen MP, Wilson MR (1989) J Computer-Aided Molec Design 3: 335
78. Frenkel D, Mulder BM, McTague JP (1984) Phys Rev Lett 52: 287
79. Frenkel D, Mulder BM (1985) Molec Phys 55: 1171
80. Frenkel D (1988) J Phys Chem 92: 3280
81. McGrother SC, Williamson DC, Jackson G (1996) J Chem Phys 104: 6755
82. Wilson MR, Allen MP (1993) Molec Phys 80: 277
83. de Miguel E, Rull LF, Chalam MK, Gubbins KE (1991) Molec Phys 74: 405
84. Luckhurst GR, Stephens RA, Phippen RW (1990) Liq Cryst 8: 451
85. Berardi R, Emerson APJ, Zannoni C (1993) J Chem Soc Faraday Trans 89: 4096
86. de Miguel E, del Rio EM, Brown JT, Allen MP (1996) J Chem Phys 105: 4234
87. Allen MP, Frenkel D (1988) Phys Rev A 37: 1813
88. Allen MP, Warren MA, Wilson MR, Sauron A, Smith W (1996) J Chem Phys 105: 2850
89. Allen MP (1990) Phys Rev Lett 65: 2881
90. Komolkin AV, Molchanov YuV, Yakutseni PP (1989) Liq Cryst 6: 39
91. Picken SJ, van Gunsteren WF, van Duijnen PTh, de Jeu WH (1989) Liq Cryst 6: 357
92. Jung B, Schürmann BL (1990) Molec Cryst Liq Cryst 185: 141
93. Ono I, Kondo S (1991) Molec Cryst Liq Cryst 8: 69
94. Ono I, Kondo S (1992) Bull Chem Soc Jpn 65: 1057
95. Ono I, Kondo S (1993) Bull Chem Soc Jpn 66: 633
96. Krömer G, Paschek D, Geiger A (1993) Ber Bunsenges Phys Chem 97: 1188

97. Huth J, Mosell T, Nicklas K, Sariban A, Brickmann J (1994) J Phys Chem 98: 7685
98. Cross CW, Fung BM (1994) J Chem Phys 101: 6839
99. Yoneya M, Berendsen HJC (1994) J Phys Soc Jpn 63: 1025
100. Kolmolkin AV, Laaksonen A, Maliniak A (1994) J Chem Phys 101: 4103
101. Yakovenko SY, Muravski AA, Krömer G, Geiger A (1995) Mol Phys 86: 1099
102. Yakovenko SY, Krömer G, Geiger A (1996) Molec Cryst Liq Cryst 275: 91
103. Hauptmann S, Mosell T, Reiling S, Brickmann J (1996) Chem Phys 208: 57
104. McBride C, Wilson MR, Howard JAK (1998) Molec Phys 93: 955
105. van der Ploeg P, Berendsen HJC (1982) J Chem Phys 76: 3271
106. van der Ploeg P, Berendsen HJC (1983) Molec Phys 49: 233
107. Egberts E, Berendsen HJC (1988) J Chem Phys 89: 3718
108. Biswas A, Schürmann BL (1991) J Chem Phys 95: 5377
109. Pastor RW (1994) Curr Opin Struct Biol 4: 486
110. Tobias DJ, Tu K, Klein ML (1997) Curr Opin Colloid & Interface Sci 2: 15
111. Cross CW, Fung BM (1995) Molec Cryst Liq Cryst 262: 507
112. Wilson MR (1997) J Chem Phys 107: 8654
113. Wilson MR (1996) GBMOL, a molecular dynamics package for modeling systems composed of Lennard-Jones and Gay-Berne particles
114. Lyulin AV, Muataz SA, Allen MP, Wilson MR, Neelov I, Allsopp NK (1998) (in press)
115. Wilson MR, Allen MP, Warren MA, Sauron A, Smith W (1997) J Comput Chem 18: 478

Computer Simulation of Liquid Crystal Phases Formed by Gay-Berne Mesogens

M. A. Bates[1], G. R. Luckhurst[2]

[1] FOM Institute for Atomic and Molecular Physics, Kruislaan 407, 1098 SJ Amsterdam,
The Netherlands
[2] Department of Chemistry and Southampton Liquid Crystal Institute,
University of Southampton, Southampton SO17 1BJ, UK

Computer simulation techniques are now widely used and accepted as an important tool to aid the understanding of the statistical mechanics of condensed matter systems. Here we review their application in the field of liquid crystals. In particular, we concentrate on the structure of the liquid crystalline phases exhibited by the family of mesogens interacting via the Gay-Berne potential, since this semi-realistic potential model has been the subject of many studies. We first describe the models used to simulate liquid crystals and how the molecular organisation in liquid crystals can be quantified. We then give a detailed discussion of the structure of the liquid crystalline phases exhibited by rod and disc shaped mesogenic molecules. The remainder of the article considers two general aspects; the enhancement of the Gay-Berne potential to include molecular features such as electrostatic and chiral interactions, flexibility and biaxiality and the use of the Gay-Berne potential to understand the behaviour of liquid crystals in complicated situations such as in mixtures and at interfaces. In each case, we give details of the model systems studied and discuss the resulting mesophase behaviour and structure.

Keywords: liquid crystals, Gay-Berne, computer simulations, phase structure

Structure and Bonding, Vol. 94
© Springer Verlag Berlin Heidelberg 1999

1
Introduction

The Monte Carlo and molecular dynamics computer simulation techniques are now well-established for determining the structure of condensed phases and these methodologies are described in several texts [1, 2]. Although the methods were originally used in the investigation of the molecular organisation in isotropic liquids they were subsequently applied to liquid crystals which are characterised by their long range orientational order and for some phases by long range positional order [3]. The presence of the long range order and the weakness of the transitions between the phases places severe demands on the computer simulation both in terms of the system size and the length of the simulation. Nonetheless computer simulations of model liquid crystals have considerably enhanced our understanding of their behaviour. The most important feature of such studies is the model potential chosen to represent the molecular interactions and various potentials varying both in complexity and reality have been proposed.

The first of these was introduced in the seminal study by Lebwohl and Lasher [4]. In their model the rod-like particles were confined to the sites of a simple cubic lattice thus avoiding the problem of specifying the scalar potential. The anisotropic potential was given a simple form depending only on the relative orientations of nearest neighbours. Despite the simple and possibly unrealistic form of the model it was found to provide a surprisingly good description of many facets of the nematic-isotropic transition. Subsequently the lattice was removed and the rod-like particles were allowed to interact through purely repulsive interactions [5]. This added some realism to the model and the structures of the liquid crystal phases which are formed show some correspondence to those of real mesogens. At the other extreme to the hard particle models are the so-called atomistic models [6] which include attractive as well as repulsive interactions. In such models the total pairwise interaction is approximated by the sum of all of the atom-atom interactions together with the torsional potentials which govern the molecular conformations. Although these models are very realistic they are especially demanding in terms of computing resources. In addition, without studying a wide range of systems it is extremely difficult to discern the relationship between the phase behaviour and the molecular structure, as for real systems.

There is then a need for a computationally tractable potential but one which includes anisotropic interactions, both attractive and repulsive, in a readily discernable manner. A solution to this problem is to use a single site or Corner potential [7] which depends on the molecular separation with the parameters in the potential being a function of the molecular orientations. One such potential is that proposed by Gay and Berne [8]; since this was first shown to exhibit a range of liquid crystal phases [9, 10] it has been the subject of many studies which have revealed the value of this potential model in understanding liquid crystal behaviour for both calamitic and discotic systems. In view of the widespread use of the Gay-Berne potential we shall confine our attention to it in this review. We begin, in the following section, with a description of the potential and discuss its parametrization. In Sect. 3 we then turn to the question of the characterisation of the phases obtained in the simulations. We digress briefly in Sect. 4 where we consider the phase behaviour of hard ellipsoids, for as we shall see they constitute a limiting case of the Gay-Berne potential. The mesophases formed by rod-like Gay-Berne molecules are described in Sect. 5 and the relationship between this phase behaviour and the parametrization of the potential is discussed. At the other extreme of anisometric shapes the phases formed by discotic Gay-Berne molecules are described in Sect. 6. It has long been known that dipolar forces can have a profound effect on liquid crystal behaviour. This has been studied using simulations by adding a dipolar interaction to the Gay-Berne potential and the phases formed by these systems are discussed in Sect. 7. This also includes a description of the influence of another electrostatic interaction on mesophase behaviour, namely the quadrupolar interaction. The introduction of chiral interactions into anisometric molecules can lead to the creation of chiral mesophases. Such behaviour has been explored for Gay-Berne molecules with the results described in Sect. 8. The flexibility of molecules normally resulting from the presence of alkyl chains is known to have a profound influence on their ability to exhibit liquid crystal phases. This important aspect of structure-property relationships for Gay-Berne mesogens is described in Sect. 9. The possibility of creating a thermotropic biaxial nematic phase is a topic of considerable interest and the ways by which the Gay-Berne potential can be modified to aid in the design of such systems is discussed in Sect. 10. The technological uses of liquid crystals rely on the creation of mixtures with the desired properties. The mesophase behaviour of binary mixtures of Gay-Berne molecules both rod-like and disc-like is considered in Sect. 11. Finally in

Sect. 12 we describe the results of simulations of Gay-Berne molecules at interfaces and in confined geometries.

2
The Gay-Berne Potential

The creation of the Gay-Berne potential owes much to the original work by Corner in his pioneering development of pair potentials for molecules [7]. He had noted that the Lennard-Jones 12–6 potential provided a good description of the interactions between atoms

$$U_{LJ}(r) = 4\varepsilon\{(\sigma/r)^{12} - (\sigma/r)^6\}; \tag{1}$$

ε being the well depth and σ the contact distance at which the repulsive and attractive interactions are balanced. He had argued, implicitly, that the same function could be used to describe the interaction between molecules provided the parameters σ and ε were allowed to vary with the molecular orientations as well as that of the intermolecular vector. He attempted to develop analytic functions for this dependence of σ and ε but they are not suitable for the large shape anisotropies necessary to observe liquid crystal behaviour where typically the length-to-breadth ratio is greater than about 3.

The way forward was proposed by Berne and Pechukas [11] many years later. Their important idea was to consider the overlap between two prolate ellipsoidal gaussian distributions. From the expression for this overlap they evaluated a range parameter which was taken to be the contact distance σ and a strength parameter which was set equal to the well depth, ϵ. If the orientations of the two rod-like molecules in the laboratory frame are represented by the unit vectors $\hat{\mathbf{u}}_i$ and $\hat{\mathbf{u}}_j$ and the orientation of the intermolecular vector by the unit vector $\hat{\mathbf{r}}$ then the expression for the angular dependence of the contact distance is

$$\sigma(\hat{\mathbf{u}}_i\hat{\mathbf{u}}_j\hat{\mathbf{r}}) = \sigma_o\left(1 - \chi\left\{\frac{(\hat{\mathbf{u}}_i \cdot \hat{\mathbf{r}})^2 + (\hat{\mathbf{u}}_j \cdot \hat{\mathbf{r}})^2 - 2\chi(\hat{\mathbf{u}}_i \cdot \hat{\mathbf{r}})(\hat{\mathbf{u}}_j \cdot \hat{\mathbf{r}})(\hat{\mathbf{u}}_i \cdot \hat{\mathbf{u}}_j)}{1 - \chi^2(\hat{\mathbf{u}}_i \cdot \hat{\mathbf{u}}_j)^2}\right\}\right)^{-1/2}. \tag{2}$$

The parameter defining the molecular anisotropy χ is given by

$$\chi = \{(\kappa^2 - 1\}/\{\kappa^2 + 1\}, \tag{3}$$

where κ is the ratio σ_e/σ_s of the contact distances σ_e and σ_s for the particles in an end-to-end arrangement and a side-by-side arrangement, respectively. It follows then that for a sphere χ vanishes while for an infinitely long rod it is unity and for an infinitely thin disc it is minus one. The scaling parameter σ_o is the side-by-side contact distance for a pair of rods. The angular dependence of the well depth is given by the much simpler expression

$$\varepsilon(\hat{\mathbf{u}}_i\hat{\mathbf{u}}_j\hat{\mathbf{r}}) = \varepsilon_o\{1 - \chi^2(\hat{\mathbf{u}}_i \cdot \hat{\mathbf{u}}_j)\}^{-1/2}. \tag{4}$$

It is of interest to note that in this model the anisotropy in the attractive energy is determined by the same parameter, χ, as that controlling the anisotropy in the repulsive energy. In these expressions for the contact distance and the well depth their angular variation is contained in the three scalar products $\hat{u}_i \cdot \hat{u}_j$, $\hat{u}_i \cdot \hat{r}$ and $\hat{u}_j \cdot \hat{r}$ which are simply the cosines of the angle between the symmetry axes of the two molecules and the angles between each molecule and the intermolecular vector.

Subsequently, it was appreciated that there are two major difficulties with this model potential. One was the observation that the width of the attractive well varied with the molecular orientations which is unrealistic [12]. Equally unrealistic is the prediction that the well depth depends only on the relative orientation of the two particles and not on their orientation with respect to the intermolecular vector (see Eq. 4). These difficulties were addressed by several groups [13] and culminated in the proposals by Gay and Berne [8] which are essentially ad hoc in character. To remove the angular variation of the width of the attractive well they changed the functional form from a dependence on the scaled distance (r/σ) (see Eq. 1) to a shifted and scaled separation R where

$$R = \{r - \sigma(\hat{u}_i\hat{u}_j\hat{r}) + \sigma_o\}/\sigma_o \tag{5}$$

thus giving a Kihara-like potential. The independence of the well depth on the molecular orientation with respect to the intermolecular vector was removed by constructing a new contribution to the well depth which has much in common with that of the contact distance, $\sigma(\hat{u}_i\hat{u}_j\hat{r})$. In fact

$$\varepsilon(\hat{u}_i\hat{u}_j\hat{r}) = \varepsilon_o\, \varepsilon''(\hat{u}_i\hat{u}_j)\varepsilon''''(\hat{u}_i\hat{u}_j\hat{r}), \tag{6}$$

where

$$\varepsilon'(\hat{u}_i\hat{u}_j\hat{r}) = 1 - \chi'\left(\frac{(\hat{u}_i.\hat{r})^2 + (\hat{u}_j.\hat{r})^2 - 2\chi'(\hat{u}_i.\hat{r})(\hat{u}_j.\hat{r})(\hat{u}_i.\hat{u}_j)}{1 - \chi'(\hat{u}_i.\hat{u}_j)^2}\right) \tag{7}$$

and

$$\chi' = (\kappa'^{1/\mu} - 1)/(\kappa'^{1/\mu} + 1). \tag{8}$$

Here, κ' provides a measure of the anisotropy in the well depth and for rod-like molecules is the ratio $\varepsilon_s/\varepsilon_e$ where ε_s and ε_e are the well depths when the molecules are side-by-side and end-to-end, respectively. The scaling parameter ε_o is the well depth when the molecules are in the cross configuration as we can see by setting $\hat{u}_i \cdot \hat{u}_j = \hat{u}_i \cdot \hat{r} = \hat{u}_j \cdot \hat{r} = 0$ in Eqs. (4), (6) and (7). We should note that in the limit κ and κ' tend to unity, that is χ and χ' vanish then the Gay-Berne potential is reduced to the Lennard-Jones 12–6 potential.

The Gay-Berne potential depends on four parameters in its scaled form, that is $U(\hat{u}_i\hat{u}_jr)/\varepsilon_o$ as a function of the scaled distance r/σ_o; these parameters are κ,

κ', μ and v. There are then an infinite number of Gay-Berne potentials depending on the values given to these parameters. It seems sensible, therefore, to introduce a mnemonic to denote these and GB(κ, κ', μ, v) has been proposed [14]. The choice of κ is relatively straightforward for it is essentially the length of the molecule divided by its width and this is readily evaluated for typical mesogenic molecules. The determination of the remaining three, κ', μ and v is far more problematic. In their original work Gay and Berne [8] took κ to be 3 which is the minimum necessary to observe liquid crystal behaviour for real systems. They then constructed an atomistic potential by placing four Lennard-Jones interaction sites in a line and mapped the Gay-Berne potential onto the interaction energy of two such arrays. This procedure gave $\kappa' = 5$, $\mu = 2$ and $v = 1$ in addition to the value of 3 for κ. Although the Gay-Berne mesogen GB(3.0, 5.0, 2, 1) has been well studied it is not clear that this set of parameters is typical of real mesogenic molecules. In addition the shape of an array of Lennard-Jones sites is essentially spherocylindrical while that of a Gay-Berne particle is ellipsoidal, as we shall see. To explore these features of the parametrization of the Gay-Berne potential, Luckhurst and Simmonds have mapped this onto an atomistic model for the interaction between two p-terphenyl molecules [15]. They found that $\kappa = 4.4$, $\kappa' = 39.6$, $\mu = 0.8$ and $v = 0.74$. It would seem that most of the parameters differ from those introduced by Gay and Berne, the most dramatic difference is for the well-depth anisotropy which is seen to be significantly larger for a typical mesogenic molecule than the value of 5 which they had proposed. We shall consider the relationship between the liquid crystallinity of the Gay-Berne mesogen GB(κ, κ', μ, v) and the values of the parameters in Sect. 5.

Although the mathematical form of the Gay-Berne potential is obviously essential for computer simulation studies it is also helpful to have a graphical representation of the potential to help understand the results of the simulations. This can be achieved in a variety of ways and the most straightforward of these is to plot the distance dependence of the potential for particular orientations of the two molecules. Such a representation is shown in Fig. 1 for GB(4.4, 20, 1, 1) [14] which is closely related to the parameter set proposed by Luckhurst and Simmonds [15] as typical of those for real mesogenic molecules. The essential constancy of the width of the attractive well as the molecular orientations are changed is in accord with what is expected for real systems as are the other features of the potential function. Thus the contact distance changes dramatically from a small value when the rod-like particles are side-by-side to a large value when they are end-to-end. In addition the depth of the attractive well also changes significantly, being large when the molecules are side-by-side and small when they are end-to-end. Similarly the attraction is weak for the side-to-end arrangement but becomes stronger when the molecules are in a cross configuration. An alternative representation of the potential is to plot the energy contours for a pair of molecules confined to a plane and with a fixed relative orientation. Clearly there are many different choices but for rod-like molecules having their symmetry axes parallel it is expected to correspond to a particularly probable arrangement, especially in a liquid crystal phase. The energy contours for this

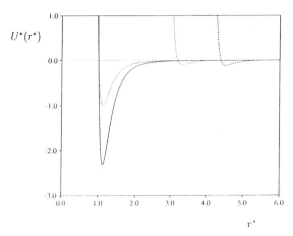

Fig. 1. The distance dependence of the Gay-Berne potential for GB(4.4, 20.0, 1,1) when the two molecules are in side-by-side (—), end-to-end (- - - - - -), side-to-end (·········) and cross (············) arrangements

arrangement are shown in Fig. 2, with the zero energy contour as the innermost. It would seem from the appearance of this that the Gay-Berne molecules have an ellipsoidal shape. Indeed this proves to be analytically correct as can be seen from the expression for the contact distance in Eq. (2) when $\hat{u}_i \cdot \hat{u}_j$ is set equal to unity. The strong anisotropy in the attractive interactions is also apparent from the contour plot with low energy contours when the molecules are side-by-side but not end-to-end. It is important to note that the shape of the molecule implicitly given by such contour plots of the energy does vary with the relative molecular orientation. For example, when the molecules are orthogonal to each other and to the intermolecular vector then the zero-energy contour again has an elliptical shape although with a different ellipticity. In marked contrast, if the molecules are orthogonal to each other but with one parallel to the intermolecular vector and the other orthogonal to it then the zero-energy contour is circular [15]. This contrasts with the essentially square shape found for real mesogenic molecules [15]; although this marked difference is clearly unsatisfactory, such a molecular arrangement is most unlikely in a liquid crystal phase.

Although the Gay-Berne potential was developed for rod-like molecules it should clearly be applicable for disc-like molecules given an appropriate choice of parameters. For rods the parameters κ and κ' have been defined in terms of the contact distance and well depth when the two molecules are in side-by-side and end-to-end arrangements. For discs the equivalent configurations are edge-to-edge and face-to-face so that κ is now σ_f/σ_e and so less than unity; κ' is $\varepsilon_e/\varepsilon_f$ and is also expected to be less than unity. The values of these ratios can be obtained by mapping the Gay-Berne potential onto an atomistic potential for a typical discotic molecule, as for rod-like particles [15]. This approach has been used for triphenylene which is the core moiety for many discotic liquid crystals and gives values for κ of 0.345 and for κ' of 1/9

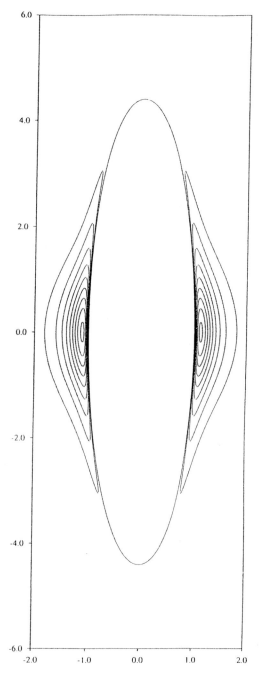

Fig. 2. The energy contours for two GB(4.4, 20.0, 1, 1) molecules confined to a plane with their symmetry axes parallel. The contours are separated by 0.25 and range from 0 to −2.25

[16]. However, the well-depth anisotropy was found to be too large and that κ' equal to 1/5 does result in liquid crystal formation [16].

There is an additional feature of the Gay-Berne potential which needs to be considered when it is used to represent the interaction between disc-like molecules. This relates to the scaled and shifted separation R between the two molecules (see Eq. 5). For rods the separation r is shifted by $(\sigma(\hat{u}_i\hat{u}_j\hat{r}) - \sigma_0)$ and scaled with σ_0; from the definition of the contact distance in Eq. (2) it follows that σ_0 is equal to σ_s which is the contact distance for the side-by-side configuration. Since the potential energy tends to infinity as R tends to zero (see Eq. 1) the separation r when this occurs is then $(\sigma(\hat{u}_i\hat{u}_j\hat{r}) - \sigma_0)$. For rods in a side-by-side arrangement the energy is infinite when r is equal to σ_s and for the end-to-end arrangement when r is equal to $\sigma_e-\sigma_s$. This is entirely reasonable because the energy should become infinite when the nuclei are in contact. If the same form for R is used for discs then for the edge-to-edge arrangement the energy is infinite when r is zero. For the face-to-face arrangement the energy tends to infinity as r tends to $\sigma_f-\sigma_e$. Clearly neither limit is realistic; indeed for the face-to-face arrangement the energy is only infinite when the separation r is negative. In order to correct for this flaw in the definition of the scaled and shifted separation R all that we need do is to identify σ_0 in the shifting term (see Eq. 5) as σ_f [17]. Then the infinite energy separation r is $\sigma_e-\sigma_f$ for the edge-to-edge configuration and zero for the face-to-face configuration as we would expect for real systems. However, in order to facilitate comparisons with simulations of rod-like molecules it is sensible to retain the distance scaling parameter as σ_e.

3
Characterisation of Phase Structure

The defining characteristic of a liquid crystal is the long range orientational order in that the molecular symmetry axis, of either rods or discs, tends to be parallel to the director. At the single molecule level the complete description of the orientational order is the singlet orientational distribution function, $f(\beta)$, which gives the probability of finding a molecule with its symmetry axis at an angle β to the director [18]. It is, clearly, straightforward to extract $f(\beta)$ from computer simulations which provide the co-ordinates of the molecules taken from the correct equilibrium distribution of configurations. In contrast it is extremely difficult to determine the singlet orientational distribution function for real mesogens. However, it is relatively easy to measure an order parameter with which to characterise the orientational order of liquid crystals. The most readily available order parameter is the average of the second Legendre polynomial

$$\bar{P}_2 = (3\overline{\cos^2\beta} - 1)/2 \tag{9}$$

which vanishes in the isotropic phase and is unity for a perfectly oriented system. This order parameter is not unique and, given the apolar symmetry of most liquid crystal phases, any even rank Legendre polynomial could be used to characterise the orientational order, although as the rank increases so too does the difficulty in measuring the order parameter. In consequence although values of the fourth rank order parameter, $\bar{P}_4$, defined by

$$\bar{P}_4 = (35\overline{\cos^4 \beta} - 30\overline{\cos^2 \beta} + 3)/8 \tag{10}$$

have been determined results for higher rank order parameters are much rarer. If the singlet orientational distribution function is known then it is possible to calculate any order parameter via

$$\bar{P}_L = \int P_L (\cos \beta) f(\beta) \sin \beta \ d\beta \tag{11}$$

while in principle all of the order parameters must be known to reconstruct the distribution function from

$$f(\beta) = \sum_L \{(2L + 1)/2\} \bar{P}_L P_L(\cos \beta), \quad L \quad \text{even.} \tag{12}$$

For smectic phases the defining characteristic is their layer structure with its one dimensional translational order parallel to the layer normal. At the single molecule level this order is completely defined by the singlet translational distribution function, $\rho(z)$, which gives the probability of finding a molecule with its centre of mass at a distance, z, from the centre of one of the layers irrespective of its orientation [19]. Just as we have seen for the orientational order it is more convenient to characterise the translational order in terms of translational order parameters τ_n which are the averages of the Chebychev polynomials, $T_n(\cos 2\pi z/d)$; for example

$$\tau_1 = \overline{\cos 2\pi z/d}, \tag{13}$$

where d is the layer spacing, that is the periodicity in the distribution function $\rho(z)$. There are very few measurements of the translational order parameters for real mesogens and those which are available are of low rank, that is τ_1 and τ_2. As for the orientational order parameters the translational order parameters can be obtained by a convolution of $\rho(z)$ with the appropriate Chebychev polynomial. Similarly the singlet translational distribution function can be expanded in a basis of Chebychev polynomials with the expansion coefficients proportional to the translational order parameters of the same rank,

$$\rho(z) = (1/d) + (2/d) \sum_{n\neq o} \tau_n T_n(\cos 2\pi z/d), \tag{14}$$

with no restriction on the value of the integer, n.

The two singlet distribution functions are not in themselves sufficient to characterise the order in a smectic A phase because there is, in general, a correlation between the position of a molecule in a smectic layer and its orientation. We need, therefore, the mixed singlet distribution function $P(z, \cos\beta)$ which gives the probability of finding a particle at position z and at an orientation β with respect to the director [18, 19]. At the level of description provided by the order parameters it is necessary to introduce the mixed order parameter

$$\sigma_{L,n} = \overline{P_L(\cos\beta)\cos(2\pi nz/d)} \tag{15}$$

to complement the pure translational and orientational order parameters. With the limit that there is no correlation between position and orientation then the mixed order parameter is simply a product of the pure order parameters [20]. The complete singlet distribution function $P(z, \cos\beta)$ is related to the order parameters which appear as part of the coefficients in an expansion of $P(z, \cos\beta)$ in a basis of Legendre and Chebychev polynomials [18, 19].

More information about the structure of a phase is contained in a hierarchy of pair distribution functions which give the probability of finding two molecules with particular positions and orientations. The simplest of these is the radial distribution function, $g(r)$, which gives the probability of finding a particle at a distance, r, from another, irrespective of their orientations. This function is normalised so that in the absence of positional correlations it tends to a value of unity. The anisotropy of liquid crystal phases means that it is valuable to introduce additional positional distribution functions to reflect this structural feature. For example, for uniaxial phases, we can define a longitudinal distribution function $g_\parallel(r_\parallel)$ where $r_\parallel$ is the component of the separation between the two particles resolved along the director. Similarly there is a transverse distribution function $g_\perp(r_\perp)$ where $r_\perp$ is the component of the separation resolved perpendicular to the director. Clearly $g_\parallel(r_\parallel)$ would be used to distinguish between a nematic and a smectic A phase, while $g_\perp(r_\perp)$ would assist in distinguishing between smectic A and B phases.

In addition to characterising the long range structure of a phase the pair distribution functions could also be used to explore the local structure. However, some care is needed in interpreting even the relatively simple radial distribution function because of its normalisation and the marked anisotropy of mesogenic molecules. Consider, for example, a system of disc-like molecules, here the probability of finding two molecules at a separation of approximately the molecular thickness is restricted to those in a face-to-face arrangement. In contrast the normalisation factor is much greater for this separation, corresponding to the number of particles which can be accommodated in a thin spherical shell of the same radius. As a consequence the peak in $g(r)$ associated with the face-to-face arrangement will appear weak and less than unity. This could be taken to suggest, quite incorrectly, that the local structure is not columnar in nature. To overcome such difficulties Veerman and Frenkel introduced a columnar distribution function, $g_c(r_\parallel)$ [21]. This gives the probability, as the name suggests, of finding a molecule in a cylinder of radius R, comparable to that of the molecule, with the cylinder axis parallel to the symmetry axis of the molecule at the origin. The distance $r_\parallel$ is the separation between the two molecules resolved along the cylinder axis. Analogous problems are also encountered with rod-like molecules in determining the local smectic-like structure. These might be overcome by defining a layer distribution function, $g_l(r_\perp)$, which would give the probability of finding a molecule within a layer of thickness D comparable to the molecular length with the layer normal parallel to the symmetry axis of the molecule at the origin. The separation between the two molecules is resolved onto the layer plane and this distance is denoted by $r_\perp$.

The anisotropy of the liquid crystal phases also means that the orientational distribution function for the intermolecular vector is of value in characterising the structure of the phase [22]. The distribution is clearly a function of both the angle, β_r, made by the intermolecular vector with the director and the separation, r, between the two molecules [23]. However, a simpler way in which to investigate the distribution of the intermolecular vector is via the distance dependent order parameters $\bar{P}_{L+}(r)$ defined as the averages of the even Legendre polynomials, $P_L(\cos\beta_r)$. As with the molecular orientational order parameters those of low rank namely $\bar{P}_2^+(r)$ and $\bar{P}_4^+(r)$, prove to be the most useful for investigating the phase structure [22].

The distribution of the intermolecular vector is also of value in distinguishing between smectic A and smectic B phases with the latter having long range bond orientational order [23, 24]. At the local level we can define a bond orientational order parameter, $\Psi_6(r_i)$ for molecule i at position r_i by [25]

$$\psi_6(r_i) = \sum_k w(r_{ik}) \exp(6i\theta_{ik}) \Big/ \sum_k w(r_{ik}). \tag{16}$$

Here the summation is over molecules k in the same smectic layer which are neighbours of i and θ_{ik} is the angle between the intermolecular vector (r_i-r_k) projected onto the plane normal to the director and a reference axis. The weighting function $w(r_{ik})$ is introduced to aid in the selection of the nearest neighbours used in the calculation of $\Psi_6(r_i)$. For example $w(r_{ik})$ might be unity for separations less than say 1.4 times the molecular width and zero for separations greater than 1.8 times the width with some interpolation between these two. The phase structure is then characterised via the bond orientational correlation function

$$g_6(r_\perp) = \overline{\psi_6(r_i)\psi_6^*(r_j)} \tag{17}$$

in which $r_\perp \equiv r_i - r_j$. In the limit of perfect correlations between the intermolecular vectors for the groups centred at r_i and r_j then $g_6(r_\perp)$ is unity. At the other extreme when the intermolecular vectors are uncorrelated at long range, as in a smectic A phase, then $g_6(r_\perp)$ vanishes.

When the system contains more than one component it is important to be able to explore the distribution of the different components both locally and at long range. One way in which this can be achieved is to evaluate the distribution function for the different species. For example in a binary mixture of components A and B there are four radial distribution functions, $g^{AA}(r)$, $g^{BB}(r)$, $g^{AB}(r)$ and $g^{BA}(r)$ which are independent under certain conditions. More importantly they would, with the usual definition, be concentration dependent even in the absence of correlations between the particles. It is convenient to remove this concentration dependence by normalising the distribution function via the concentrations of the components [26]. Thus the radial distribution function of $g^{AB}(r)$ which gives the probability of finding a molecule of type B given one of type A at the origin is obtained from

$$g^{AB}(r) = n^{AB}(r)V/N_A N_B 4\pi r^2 \delta r \ , \tag{18}$$

where $n^{AB}(r)$ is the average number of molecules separated by distances between $r-(1/2)\delta r$ and $r+(1/2)\delta r$, V is the volume of the system, N_A and N_B are the number of molecules of type A and type B, respectively [27].

We turn now to the orientational correlations which are of particular relevance for liquid crystals; that is involving the orientations of the molecules with each other, with the vector joining them and with the director [17, 28]. In principal they can be characterised by a pair distribution function but in view of the large number of orientational coordinates the representation of the multi-dimensional distribution can be rather difficult. An alternative is to use distance dependent orientational correlation coefficients which are related to the coefficients in an expansion of the distribution function in an appropriate basis set [17, 28].

At the simplest level the orientational correlation of molecular pairs can be characterised by the averages of the even Legendre polynomials $P_L(\cos \beta_{ij})$ where β_{ij} is the angle between the symmetry axes of molecules i and j separated by a distance r. This correlation coefficient is denoted by

$$G_L(r) = \overline{P_L(\cos \beta_{ij})} \ ; \tag{19}$$

which is unity for perfect orientational correlations. At the other limit when the direct correlations are lost, which is expected to occur for large separations,

$$\lim_{r \to \infty} G_L(r) = \overline{P_L}^2 \ . \tag{20}$$

It is possible, therefore, to use the $G_L(r)$, normally for $L = 2$ and possibly 4, to explore the extent of the local structure determined by the direct orientational correlations. An alternative and more detailed way in which to investigate the orientational correlations involving molecular pairs is to use the correlation coefficients defined by [29]

$$g_{L_iL_jm}(r) = 4\pi g(r)\,\overline{Y^*_{L_im}(\omega_i)\,Y^*_{L_j-m}(\omega_j)}\ . \tag{21}$$

Here, $Y_{L_i}m_{(\omega_i)}$ denotes a spherical harmonic, ω_i represents the spherical polar angles made by the symmetry axis of molecule i in a frame containing the intermolecular vector as the z axis. The choice of the x and y axes is arbitrary because the product of the functions being averaged depends on the difference of the azimuthal angles for the two molecules which are separated by distance r. At the second rank level the independent correlation coefficients are

$$g_{020}(r) = \sqrt{5}g(r)\overline{(3\cos^2\beta_j - 1)}/2\ ,$$
$$g_{220}(r) = 5g(r)\overline{\{(3\cos^2\beta_i - 1)/2\}\{(3\cos^2\beta_j - 1)/2\}}\ ,$$
$$g_{221}(r) = (15/2)g(r)\overline{\sin\beta_i\cos\beta_i\sin\beta_j\cos\beta_j\exp\ i(\alpha_i - \alpha_j)}\ ,$$
$$g_{222}(r) = (15/8)g(r)\overline{\sin^2\beta_i\sin^2\beta_j\exp 2i(\alpha_i - \alpha_j)} \tag{22}$$

and values of these can be used to aid in the determination of the local molecular organisation [14]. At long range when the orientational correlations are lost the coefficients adopt particularly simple forms

$$\lim_{r\to\infty}\ g_{L_iL_jm}(r) = (-)^m g(r)\delta_{L_iL_j}\overline{P_{L_i}}\,\overline{P_{L_j}}\ , \tag{23}$$

and in this limit it is also to be expected that the radial distribution function will have reached its limiting value of unity.

A knowledge of the order parameters, singlet distribution functions, correlation coefficients and pair distribution functions clearly provides a quantitative route to the identification of a phase and the characterisations of its structure. There is, however, a complementary, essentially qualitative, method with which to assign a phase structure and this is simply to view the molecular organisation directly by constructing an image of a configuration using the coordinates taken from the production stage of the simulation. The image can be viewed in a variety of ways so allowing the long range structure to be clearly identified; this should normally result in an unambiguous identification of the liquid crystal phase given their manifest differences. However, the snapshot of the phase obtained in this way only provides an instantaneous picture of the molecular organisation in the phase. This will certainly vary during the simulation and will differ from the average structure. Nonetheless it seems unlikely that such images obtained from a single configuration will provide an incorrect identification of a liquid crystal phase except, of course, when there are subtle differences between the phases. As we shall see the visualisation of configurations taken from simulations provides a valuable guide to the behaviour of a model system at the molecular level.

The identification of the phases formed by real mesogens can be achieved in a variety of ways but X-ray diffraction proves to be a powerful and rigorous technique [30]. In addition it can provide a method with which to characterise the order, both orientational and positional, of the phase. Given the wealth of experience in relating the diffraction pattern of a phase to its structure it seems likely than an analogous approach should be valuable in the characterisation of the structure of phases formed by model systems. The information content of a diffraction or scattering experiment is enhanced when the system studied exists as a monodomain which usually proves to be the case for the systems studied by simulations. The total coherent scattering intensity, $I_T(Q)$, is determined as a function of the scattering vector Q and for an atomistic system can be calculated in the following way. The total scattering amplitude is given by the sum over all N_a atoms as [31]

$$F_T(Q) = \sum_{j=1}^{N_a} a_j(Q) \ \exp(i\mathbf{Q}.\mathbf{r}_j), \tag{24}$$

where $a_j(Q)$ is the atomic scattering factor for atom j at position r_j. The scattering intensity $I_T(Q)$ is then the product of the scattering amplitude with its complex conjugate

$$I_T(Q) = F_T(Q) \, F_T^*(Q) \,. \tag{25}$$

This total intensity is made up of single molecule scattering

$$I_S(Q) = F_S(Q) F_S^*(Q) \,, \tag{26}$$

where by analogy to Eq. (24)

$$F_S(Q) = \sum_{j=1}^{N_k} a_j(Q) \ \exp(i\mathbf{Q}.\mathbf{r}_j) \tag{27}$$

but now the summation is restricted to the N_k atoms in molecule k. The interference term coming from scattering between molecules is simply the difference

$$I_I(Q) = I_T(Q) - I_S(Q) \,; \tag{28}$$

and clearly both contributions are readily available from a simulation.

The implementation of this formalism for an atomistic model does not present any difficulties, at least conceptually. However, it is not obvious how the formalism should be used for single site potentials such as the Gay-Berne model. One approach would be to superimpose a mesogenic structure on the Gay-Berne ellipsoid but this structure would necessarily have a lower symmetry. Alternatively it would be possible to place a line of atoms along the symmetry axis of the ellipsoid which is more in keeping with the origins of the Gay-Berne potential [32]. However, to avoid spurious peaks in the scattering pattern a relatively large number of atoms would need to be used, giving the molecule a spherocylindrical shape. To avoid this apparent inconsistency with the Gay-Berne model, each molecule can be treated as a uniform scattering ellipsoid. The scattering factor for an ellipsoid depends on its anisotropy as well as its orientation [31, 33];

$$a_k(\hat{\mathbf{u}}_k, \mathbf{Q}^*) = (3 \sin \gamma_k - \gamma_k \cos \gamma_k)/\gamma_k^3, \tag{29}$$

where Q^* is the scaled scattering vector $Q_\alpha^* = 2\pi/r_\alpha^*$

$$\gamma_k = (1/2)|\mathbf{Q}^*|\left((\sigma_e/\sigma_s)^2 \cos^2 \varphi_k + \sin^2 \varphi_k\right)^{1/2}, \tag{30}$$

and the angle φ_k is given by

$$\cos \varphi_k = (\mathbf{Q}^* . \hat{u}_k)/|\mathbf{Q}^*| . \tag{31}$$

The ellipsoidal equivalent of the scattering amplitude for a system of atoms (see Eq. 24) is

$$F_T(\mathbf{Q}^*) = \sum_{k=1}^{N} a_k(\hat{\mathbf{u}}_k, \mathbf{Q}^*) \exp(i\mathbf{Q}^* \cdot \mathbf{r}_k^*), \tag{32}$$

where the summation is over the N molecules. The total and single molecule scattering intensities are then evaluated from Eqs. (25) and (26) using this expression for $F_T(Q^*)$ and the analogous term for $F_S(Q^*)$ [32, 33]. The calculation of the scattering patterns from simulation data requires care if artefacts resulting from the small system size are to be avoided [33]. As we shall see in Sect. 5 this does prove to be possible although the number of particles is larger than that usually used in simulation studies of Gay-Berne model mesogens.

4
Hard Ellipsoids

It is now well-established that for atomic fluids, far from the critical point, the atomic organisation is dictated by the repulsive forces while the longer range attractive forces serve to maintain the high density [34]. The investigation of systems of hard spheres can therefore be used as simple models for atomic systems; they also serve as a basis for a thermodynamic perturbation analysis to introduce the attractive forces in a van der Waals-like approach [35]. In consequence it is to be expected that the anisotropic repulsive forces would be responsible for the structure of liquid crystal phases and numerous simulation studies of hard objects have been undertaken to explore this possibility [36].

The first of these was by Vieillard-Baron [5] who investigated a system of spherocylinders but failed to detect a liquid crystal phase primarily because the anisometry, L/D, of 2 was too small [37]. He also attempted to study a system of 2392 particles with the larger L/D of 5 but these simulations had to be abandoned because of their large computational cost. However, in view of the ellipsoidal shape of the Gay-Berne particles it is the behaviour of hard ellipsoids of revolution which is of primary relevance to us.

The phase behaviour of systems of such hard ellipsoids has been studied in considerable detail by Frenkel et al. [38] for both rod-like and disc-like particles with varying degrees of ellipticity. The phase diagram resulting from these studies is shown in Fig. 3a where the density which is the controlling variable is expressed as the packing fraction. For spheres, with an ellipticity of unity, we find crystal, plastic crystal and fluid phases. As the ellipticity is increased to give a prolate ellipsoid the plastic crystal phase is found to vanish and for values greater than about 3 a nematic phase appears. As the ellipticity increases so does the stability of the nematic phase and the nematic-isotropic phase boundary approaches the Onsager limit [38]. Analogous behaviour is observed for hard oblate ellipsoids; indeed using the scales in Fig. 3a the phase behaviour exhibits an elegant symmetry for prolate and oblate ellipsoids. This is lost, however, if the density is expressed as the conventional number density used in most simulation studies of Gay-Berne model mesogens, as the plot in Fig. 3b shows. It is of interest that with these scales the region of ellipticity and number density over which nematic phases are formed by prolate hard ellipsoids appears significantly smaller than for oblate ellipsoids. Although the form of the phase diagram in Fig. 3 is not in doubt there is a dispute about the minimum ellipticity needed to observe calamitic nematic behaviour. Thus simulations by Zarragoicoechea et al. [39] have shown that the nematic phase found by Frenkel et al. [38] is not observed at the same densities when the system size is increased.

What is surprising, at least at first sight, is that systems of hard prolate ellipsoids do not form smectic phases and oblate ellipsoids do not exhibit columnar phases. This contrasts with the formation of smectic phases by hard spherocylinders [37] and of columnar phases by hard cut-spheres [40]. However, the absence of liquid crystal phases with some translational order has been explained by Frenkel [41] in the following way. In these phases the orientational order is high and so the molecular symmetry axes can be taken to be parallel to the director. Then scaling the coordinates along this direction can be accomplished so as to convert the ellipsoids into spheres without changing the configurational partition function. Since systems of hard spheres do not form columnar or layered structures it follows that hard ellipsoids of revolution should not either.

It is clear that systems of hard ellipsoids exhibit an intriguingly simple phase behaviour with some resemblance to that of real nematogens. However, such systems cannot form smectic or columnar phases and in addition the phase transitions are not thermally driven as they are for real mesogens. As we shall see in the following sections the Gay-Berne potential with its anisotropic repulsive and attractive forces is able to overcome both of these limitations.

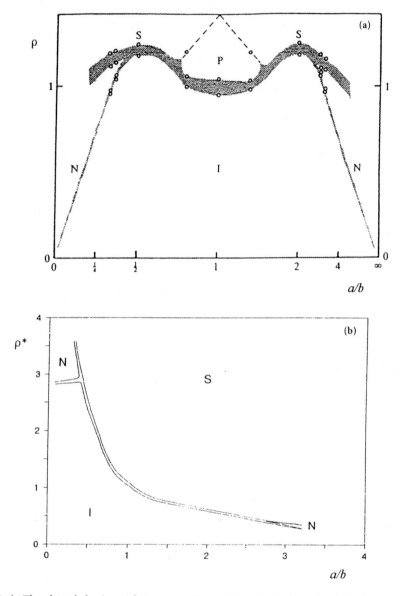

Fig. 3a,b. The phase behaviour of a system of hard ellipsoids, both prolate and oblate, as a function of the ellipticity, *a/b*, plotted against: **a** the packing fraction, ρ; **b** the scaled number density, ρ^*

5
Calamitic Gay-Berne Mesogens

One of the primary features of the Gay-Berne potential is the presence of anisotropic attractive forces which should allow the observation of thermally driven phase transitions and this has proved to be the case. Thus using the parametrisation proposed by Gay and Berne, Adams et al. [9] showed that GB(3.0, 5.0, 2, 1) exhibits both nematic and isotropic phases on varying the temperature at constant density. This was chosen to be close to the transitional density for hard ellipsoids with the same ellipticity; indeed it is generally the case that to observe a nematic-isotropic transition for Gay-Berne mesogens the density should be set in this way. The long range orientational order of the phase was established from the non-zero values of the orientational correlation coefficient, $G_2(r)$, at large separations and the translational disorder was apparent from the radial distribution function.

Although no other liquid crystal phases were reported, another important feature of the Gay-Berne potential is the anisotropy of the attractive forces which should favour the formation of smectic phases. These were discovered [10] a few years after the first simulation in an analogous study of the mesogen GB(3.0, 5.0, 1, 2) in which the values of the exponents μ and ν in the well depth function have been exchanged (see Eq. 6). This reversal does not change the ratio of the well depths for the side-to-side and end-to-end arrangements, although the side-by-side well depth is now significantly greater than the cross and T configurations [10]. Such changes should favour liquid crystal formation. Molecular dynamics simulations at constant density reveal the formation of isotropic, nematic, smectic A, smectic B and crystal phases as the temperature is lowered. These phases were identified by visualising snapshots of configurations taken from the production stage of the simulation and a selection for the five phases is shown in Fig. 4. The identification of the smectic A phase at a temperature of 1.49 is somewhat difficult because the layer structure is only just apparent from the snapshot. In addition its presence may be an artefact resulting from the relatively small system size; indeed entirely analogous effects are observed with the orientational order where in the isotropic phase obtained in simulations the order parameter $\overline{P_2}$ is non-zero. In fact a more extensive simulation of GB(3.0, 5.0, 1, 2) using a larger system and allowing the dimensions of the simulation box to change while keeping its volume fixed have failed to locate a smectic A phase [42]. The layer structure of the smectic B phase is clearly apparent at the lower scaled temperature of 1.00. However, the layer normal for the smectic B is usually found to be aligned along the main diagonal of the box and the scaled layer spacing is about 2.6 which is significantly smaller than the scaled molecular length of 3.0. It has been suggested [10] that these two features of the simulation may result from the periodic boundary conditions which together with the spatial inhomogeneity of the smectic phase could result in the periodic images not being commensurate with the particles in the simulation box. In consequence this would create defects in the structure at the edge of the box which would

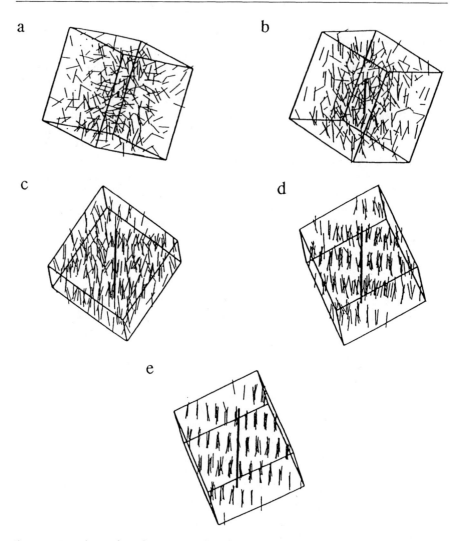

Fig. 4a–e. Snapshots of configurations taken from the production stages of simulations of GB(3.0, 5.0, 1, 2) at scaled temperatures: **a** 3.00; **b** 2.19; **c** 1.49; **d** 1.00; **e** 0.50. The molecules are represented by lines which are shorter than the molecular length; the *thick lines* show the director and their lengths are proportional to the orientational order parameter, $\overline{P_2}$, for the configuration

increase the energy of the system. The additional energy would be removed if the natural layer spacing d and the box dimension L satisfied the constraint

$$d = L/(N_x^2 + N_y^2 + N_z^2)^{1/2}, \tag{33}$$

where N_α is an integer associated with the box axis α and defining the orientation of the layer normal [10]. This constraint is most likely to be satisfied when the layer normal is along the main diagonal of the box. Clearly the constraint can also be met by changing the smectic periodicity and this may account for the difference between the small value of d^* in comparison with the molecular length.

To see whether this is the case or whether the natural structure of the smectic B phase is interdigitated it is necessary to ensure that the structure in the simulation box is commensurate with the periodic images. This can be achieved by performing Monte Carlo simulations at constant pressure which allows the dimensions of the simulation box to change [2], so ensuring that an integral number of layers can be accommodated within it. Such simulations show that the scaled smectic periodicity obtained from the singlet translation distribution function, $\rho(z)$, is indeed 2.6 [43] which means that the structure is interdigitated and that the simulations at constant volume did give the natural structure. The ability of the molecules to penetrate adjacent layers presumably results from the ellipsoidal molecular shape. It is significant that this interdigitated structure is consistent with that of real mesogens where the molecular length is often somewhat larger than the smectic periodicity [30]. This study also revealed a lack of correlation between the positions of the molecules in one layer with those in adjacent layers which supports the assignment of the phase as a smectic B rather than a crystal B. It is also of interest that at a constant scaled pressure of 1, the smectic B phase undergoes a strong first order transition directly to the isotropic phase.

A more detailed study of the phase behaviour of the mesogen GB (3.0, 5.0, 2, 1) has been undertaken by de Miguel et al. [44] as a function of both density and temperature. In addition to the isotropic, nematic and smectic B phases already observed for Gay-Berne mesogens they also studied the vapour phase and reported a tilted smectic B phase. It is difficult to see how the potential model could result in the formation of a tilted structure; however it may be significant that the smectic B phase found in the constant pressure Monte Carlo simulations appeared to have a slight rippled structure [43]. The phase diagram established by de Miguel et al. is shown in Fig. 5. It is significant that, although there is an isotropic-vapour phase boundary as well as one between the smectic B and vapour phases, there is no common boundary between the nematic and vapour phases. The absence of the nematic-vapour phase boundary clearly limits the use of this particular Gay-Berne model in the study of the surface structure of the nematic phase. To overcome this problem de Miguel et al. [45] have explored the influence of the well depth anisotropy parameter κ' on the phase behaviour while keeping the other Gay-Berne parameters fixed at their original values. They studied the mesogens GB(3.0, κ', 2, 1), where κ' was assigned the values 1, 1.25, 2.5, 5, 10 and 20, by keeping the temperature fixed, at 0.7, and varying the density. For κ' greater than, or equal to, 5 the mesogen exists as a smectic B phase which undergoes a transition directly to the isotropic phase. Such behaviour is perhaps to be expected for large values of κ' which favour the side-by-side configuration. For values of κ' smaller than 5 a nematic phase appears in the phase diagram although for κ' of 2.50 the nematic range is short and followed by a smectic B phase. In addition it seems likely that a nematic-vapour boundary appears when κ' is either 1.0 or 1.25, although the smectic B phase is not expected for these values of κ'.

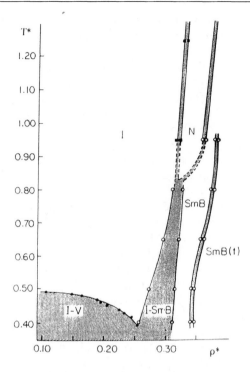

Fig. 5. The essential form of the phase diagram for the mesogen GB(3.0, 5.0, 2, 1); the *open circles* indicate the approximate coexistence lines and the *solid circles* show the density of the isotropic liquid in equilibrium with the vapour phase

The ad hoc character of the Gay-Berne potential means that in principle the parameters occurring in it can be chosen to produce any particular set of properties, as we have just seen. However, it seems sensible to see if these parameters are typical of real mesogenic molecules. Indeed one of the strengths of the potential is that some, at least, of the parameters have a clear physical interpretation. For example, κ is determined by the molecular dimensions and κ of 3 is certainly typical of mesogenic molecules although somewhat close to the minimum necessary for liquid crystal behaviour. It is not possible to assign a value to κ' in such a simple way. For rod-like molecules κ' should be greater than unity and in a sense it is expected to be related to the number of atoms in the molecule. This occurs because in the side-by-side configuration there will be interactions between all of these while in the end-to-end arrangement only a few atoms will interact. The situation is still worse for the exponents μ and ν where it is impossible to give any qualitative arguments about their values. Gay and Berne [8] overcame these difficulties by fitting their potential to that resulting from the interaction between molecules composed of four Lennard-Jones centres arranged in a line. Such hypothetical structures are not expected to be mesogenic. This observation prompted Luckhurst and Simmonds [46] to fit the Gay-Berne potential to that between p-terphenyl molecules constructed from site-site interactions. This molecule has the advantage of structural simplicity and although it does not form liquid crystal phases, at least at atmospheric pressure, it does have a virtual nematic-isotropic transition of 360 K. The fitting process gave κ as 4.4, κ' as 39.6, μ as 0.8 and ν as 0.74 [46] which represent significant differences to the values originally proposed.

Preliminary simulations for this mesogen, GB(4.4, 39.6, 0.8, 0.74), were performed as a function of temperature at a scaled number density of 0.16 which is close to the transitional density for hard ellipsoids with the same ellipticity. Based on the behaviour of the radial distribution function, $g(r)$, the orientational correlation coefficient, $G_2(r)$, and the singlet translational distribution function, $\rho(z)$, the mesogen was found to form isotropic, nematic and smectic A phases [46]. The unambiguous observation of a smectic A phase for this Gay-Berne mesogen is clearly of considerable interest, especially as other parametrisations have resulted in smectic B phases. To explore the origins of the smectic A phases simulations were performed with smaller values of the well depth anisotropy parameter, κ'. It was found that even for κ' of 5 a smectic A phase was still formed which suggests its appearance is a consequence of the larger shape anisotropy. However, when the well depth anisotropy is removed by setting κ' to unity the smectic A phase necessarily vanishes, confirming the involvement of both κ and κ' in stabilising the smectic A.

The observation of the smectic A phase has prompted Bates and Luckhurst [14] to undertake a more detailed investigation of an analogous Gay-Berne mesogen, namely GB(4.4, 20.0, 1, 1). Thus the length-to-breadth parameter was kept at the same value, κ' was reduced to 20 but this is not expected to change the phase behaviour [46] while the exponents have been given the nearest integer value of unity. This choice of parameters has been shown [47] to provide a good description of the thermodynamic quantity [48]

$$\Gamma = -(\partial \ln T / \partial \ln V)_{\overline{P_2}} \tag{34}$$

whose prediction provides a challenging test of models of nematics. Despite this success we should note that Gay-Berne mesogens are generally found to have a volume change at the nematic-isotropic transition which is about ten times larger than that for real mesogens [14]. This particular failure of the potential may well result from its ellipsoidal shape. The phase behaviour of GB(4.4, 20.0, 1, 1) was investigated at constant pressure and revealed that at low pressures the system forms isotropic, smectic A and smectic B phases. At higher pressures a nematic phase appears, in keeping with the prediction of the Clapeyron equation. In order to investigate the structure of the four phases, isotropic, nematic, smectic A and smectic B in detail a system of 16,000 molecules was investigated using canonical molecular dynamics simulations; the use of such a high number of molecules means that correlations can be studied over significantly large distances.

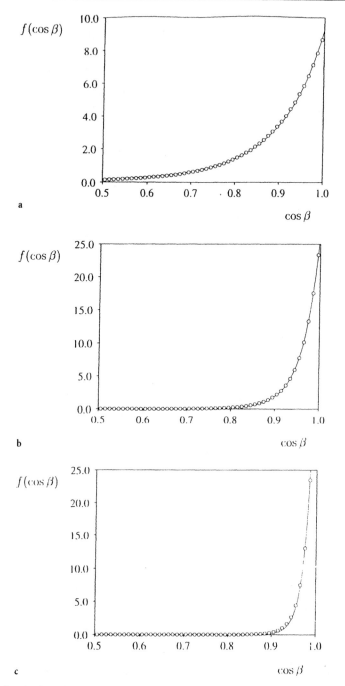

Fig. 6a–c. The singlet orientational distribution function, $f(\beta)$, for the mesogen GB(4.4, 20.0, 1, 1) in: **a** the nematic phase; **b** the smectic A phase; **c** the smectic B phase. The *solid lines* show the form of the distribution function predicted by Eq. (35)

We begin, however, with the singlet orientational distribution function which is shown for the three liquid crystal phases in Fig. 6. In each phase the distribution is peaked at $\cos\beta$ of 1 showing that the preferred molecular orientation is parallel to the director. The form of the distribution function is well represented by the relatively simple function

$$f(\beta) = Z^{-1} \exp[X\, P_2(\cos\beta)] , \qquad (35)$$

where Z is the normalisation factor, as found for GB(3.0, 5.0, 2, 1) [22]. This is in agreement with the prediction of the Maier-Saupe theory for the nematic phase [49]. However this should clearly not be taken as support for their assumed importance of anisotropic dispersion forces in stabilising nematic phases; rather it is a consequence of the dominance of the second rank order parameter $\overline{P_2}$ [50]. The reason for the success of the simple distribution function in the two smectic phases is less apparent for here the orientational order is large and so the higher rank order parameters are not negligible. It might be expected, therefore, that the distribution function would depend on all of the Legendre polynomials

$$f(\beta) = Z^{-1} \exp\left[\sum_{L(\text{even})} a_L P_L(\cos\beta) \right] , \qquad (36)$$

but in this high order limit $f(\beta)$ is only significant for β close to zero. This means that each Legendre polynomial can be expanded and the series truncated at the first non-trivial term which gives [51]

$$f(\beta) = Z^{-1} \exp(-\lambda_2 \beta^2). \qquad (37)$$

However, since the same function is obtained from the simple distribution in Eq. (35) as for the complete orientational distribution in Eq. (36) we see that even in the high order limit the Maier-Saupe form will still give a good fit to the simulation result.

Similar behaviour is found for the singlet translational distribution function, $\rho(z)$, in the two smectic phases. According to the McMillan theory [52] for smectic A phases and the variational derivation of it by Kventsel et al. [53], the distribution should take the form

$$\rho(z) = Z_\tau^{-1} \exp(g_1 \cos 2\pi z/d), \qquad (38)$$

where the strength parameter depends on both the orientational and translational order parameters. The agreement between this prediction and the simulated results is as impressive as for the orientational distribution function and for essentially analogous reasons.

The correlation between the translational and orientational order is reflected by the mixed singlet orientational and translational distribution function $P(z, \cos\beta)$. The results for this are shown in Fig. 7 for the smectic A

(a)

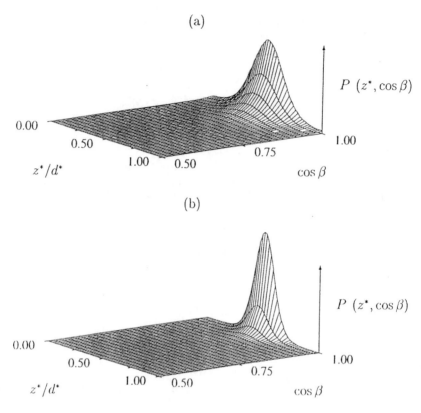

(b)

Fig. 7a,b. The singlet orientational-translational distribution function P(z*, cos β) for GB(4.4, 20.0, 1, 1) in: **a** the smectic A phase; **b** the smectic B phase

and smectic B phases. The maxima in the two distributions are clearly when the molecules are in the centre of the smectic layer and when the molecule is parallel to the director, which is to be expected. What is unexpected is that the orientational distribution function for a given position in the layer has the same form, independent of z. In other words there is no correlation between the translational and orientational coordinates as assumed by Kventsel et al. [53] but in disagreement with the McMillan theory [52]. The behaviour of P(z*, cos β) in the smectic A phase shows that in between the layers the probability of finding a molecule parallel to the director is a maximum and decreases with increasing β. This contrasts dramatically with that found for the smectic A phase formed by hard spherocylinders [54] where the orientational distribution function is bimodal with maximum probabilities when the molecule is parallel and perpendicular to the director.

One of the key quantities for a smectic phase is the scaled layer spacing, d^*, and this was determined from the periodicity in the longitudinal distribution function $g_{\parallel}(r_{\parallel})$. This distribution is unattenuated over the distances available from the simulation for the two smectic phases and the periodicity is the same although the width of the distribution is smaller in the smectic B than the smectic A phase. The value of d^* found from $g_{\parallel}(r_{\parallel})$ is 3.84 which is significantly shorter than the molecular length. The ratio of d^*/l^* is 0.88 which is very similar to that found for GB(3.0, 5.0, 2, 1) and presumably results from the ellipsoidal shape of the molecules which allows significant interdigitation. It has been suggested for real mesogenic systems that values of d/l less than unity could result from orientational fluctuations within the smectic layer. However, for the smectic A and B phases formed by the mesogen GB(4.4, 20.0, 1, 1) the orientational order is very high with $\overline{P_2}$ equal to 0.884 and 0.946, respectively and so the reduction in d^*/l^* from unity must result from interdigitation between the layers.

It is just as important to be able to distinguish between smectic A and B phases in a simulation as it is for real mesogens. For simulations this can be achieved with the aid of the bond orientational correlation function, $g_6(r_\perp)$ (see Eq. 17) and results for the two smectic phases of GB(4.4, 20.0, 1, 1) are shown in Fig. 8. They are clearly dramatically different; in the smectic A phase the correlation function decays quickly to zero indicating the absence of long range bond orientational order, in accord with the identification of this as a smectic A. In contrast within the smectic B phase $g_6(r_\perp)$ decays to a non-zero value of about 0.5, thus demonstrating the long range, but incomplete, bond orientational order of a smectic B phase, again in agreement with the identification of this phase. There is some structure on the correlation function for both the smectic phases and the most striking feature is the minima, which occur at distances corresponding to separations between shells of neighbours where the number of molecules is small. Presumably, the local positional order around a molecule at such separations is different to that in the preferred structure.

Complementary information about the structure of the smectic phases is contained in the X-ray scattering patterns, as in studies of real mesogens. The intermolecular scattering patterns calculated (see Sect. 3) for GB(4.4, 20.0, 1, 1)

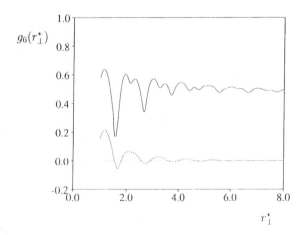

Fig. 8. The distance dependence of the bond orientational correlation function $g_6(r_\perp^*)$ found for the mesogen GB(4.4, 20.0, 1, 1) in the smectic A (- - - - -) and the smectic B (—) phases

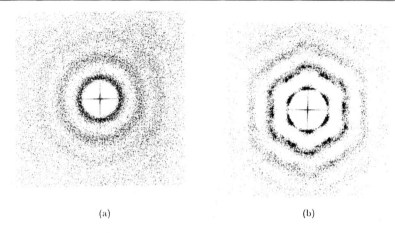

Fig. 9a,b. The intermolecular scattering patterns calculated for: **a** the smectic A; **b** the smectic B phases of the mesogen GB(4.4, 20.0, 1, 1) with the scattering vector parallel to the layer normal. The scaled scattering vectors Q_x^* and Q_y^* range from -8π to 8π

in its two smectic phases with the scattering vector parallel to the layer normal are shown in Fig. 9 [33, 55]. For the smectic A phase there is a strong first order circular pattern confirming the uniaxial nature of the phase. Weak higher order reflections are also observed which indicate some longer range translational correlations. The comparable scattering pattern for the smectic B phase (see Fig. 9b) is quite different. It contains six strong first order peaks separated from each other by a rotation of 60°. This structure confirms the existence of bond orientational order within the layers as well as a correlation between the layers so that the bond order is aligned in the same direction in different layers. There is a second ring of six diffraction peaks rotated by 30° from those in the first ring. Such peaks confirm that the local structure in the phase is hexagonal. Slightly more higher order diffractions are apparent in the smectic B phase than in the smectic A which is in accord with the anticipated longer range translational order.

The scaled pressure $P^*(\equiv P\sigma_0^3\varepsilon_0)$ used in these molecular dynamics simulations was approximately 2.0 but it is clearly of some importance to know what absolute pressure this corresponds to so that we can judge whether the conditions under which a Gay-Berne mesogen exhibits liquid crystal phases are realistic. It is necessary, therefore, to obtain estimates of the scaling parameters σ_0 and ε_0; this can be achieved in the following way [43]. The contact distance, σ_0, can be determined from X-ray studies of nematics which give the molecular width of essentially all rod-like systems to be approximately 4.5 Å. The well depth parameter, ε_0, is obtained by comparing the scaled nematic-isotropic transition temperature with that of a typical nematogen. The relatively rigid, rod-like molecular structure of 4,4′-dimethoxyazoxybenzene suggests that this could be used for such a comparison, especially as this compound has been well studied experimentally. At 1 atm the nematic-isotropic transition temperature is 409 K while T_{NI}^* obtained from the simulation at a scaled pressure of 2.0 is 1.53 which gives $N_A \varepsilon_0$ as 2.01 kJ mol^{-1}. Using this value a scaled pressure of 2.0 is found to correspond to the higher absolute pressure of 730 atm. However, since T_{NI} varies with pressure it is necessary to allow for this in the determination of ε_0; the value of dT_{NI}/dP is known for 4,4'-dimethoxyazoxybenzene and an iterative calculation gives $N_A \varepsilon_0$ as 2.19 kJ mol^{-1}. This means that the scaled pressure used in the simulation is equivalent to an absolute pressure of 790 atm which although large is not unrealistic; for example at this pressure T_{NI} for the nematogen would be 445 K which is not significantly larger than the value of 409 K at 1 atm.

6
Discotic Gay-Berne Mesogens

The prime requirement for the formation of a thermotropic liquid crystal is an anisotropy in the molecular shape. It is to be expected, therefore, that disc-like molecules as well as rod-like molecules should exhibit liquid crystal behaviour. Indeed this possibility was appreciated many years ago by Vörlander [56] although it was not until relatively recently that the first examples of discotic liquid crystals were reported by Chandrasekhar et al. [57]. It is now recognised that discotic molecules can form a variety of columnar mesophases as well as nematic and chiral nematic phases [58].

There have been several simulations of discotic liquid crystals based on hard ellipsoids [41], infinitely thin platelets [59, 60] and cut-spheres [40]. The Gay-Berne potential model was then used to simulate the behaviour of discotic systems by Emerson et al. [16] in order to introduce anisotropic attractive forces. In this model the scaled and shifted separation R (see Eq. 5) was given by

$$R = \left\{ r - \sigma(\hat{u}_i \hat{u}_j \hat{r}) + \sigma_e \right\} / \sigma_e, \qquad (39)$$

where the shifting parameter σ_e is the contact distance for the edge-to-edge configuration. The use of this shifting parameter means that the limiting form of the potential is not physically realistic when the energy is highly repulsive, although necessarily such configurations are not very probable [17]. The parameters used in the potential were obtained partly by mapping the Gay-Berne potential onto that for triphenylene and partly by exploratory simulations. The values were $\kappa = 0.345$, $\kappa' = 0.2$, $\mu = 1$ and $\nu = 2$ and so the mnemonic for this is GB(0.345, 0.2, 1, 2) where the fact that the first number is less than unity indicates the discotic character of the molecules. The distance dependence of the potential for a selection of molecular orientations is shown in Fig. 10a; the strong anisotropies in the well depth and the repulsive forces are clearly apparent.

A system of particles interacting in this way was studied using a microcanonical ensemble at a scaled density of 3.0 which is close to the transitional density for hard oblate ellipsoids with the same ellipticity (see Fig. 3). At a scaled density of 3.0 the system is found to exhibit isotropic,

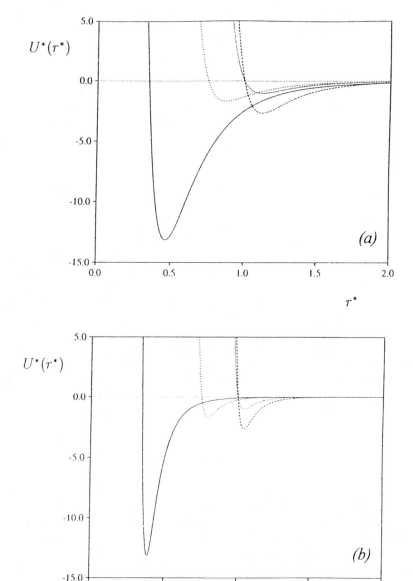

Fig. 10a,b. The distance dependence of the Gay-Berne potential for disc-like molecules calculated for a variety of orientations (——) face-to-face, (- - - - - -) edge-to-edge, (······) cross and (··········) edge-to-face with a distance shifting parameter: **a** σ_e; **b** σ_f

nematic and columnar phases. The nematic phase was identified from the unstructured form of the radial distribution function and the existence of long range orientational order. Perhaps of greater interest is the columnar phase and its structure. The phase itself is readily identified from a snapshot of a configuration taken from the production run. This also revealed that the columns have a quasi-rectangular ordering which would be expected if the molecules are tilted with respect to the column axis [3, 58] but this is not the case for this Gay-Berne discotic. An explanation is hinted at from the appearance of the longitudinal distribution function $g_\parallel(r_\parallel)$ which is shown in Fig. 11 for the columnar and nematic phases. The distribution function is essentially featureless within the nematic phase but exhibits a pronounced oscillation in the columnar phase with a scaled periodicity of about 0.25 which is just half of the distance corresponding to the minimum in the potential when the molecules are face-to-face (see Fig. 10a). Clearly the correlations between molecules at separations of 0.25 cannot originate from those in the same column but must come from those in neighbouring columns. In other words the molecules from one column are interdigitated with those of its neighbours; this is reminiscent of the interdigitation of Gay-Berne molecules in the layers of the smectic phases which we encountered in Sect. 5. The extent of the interdigitation between columns can be judged from the transverse distribution function $g_\perp(r_\perp^*)$ and is found to be extensive. In view of this significant interpenetration it is not possible to pack more than four columns around the central column which accounts for the quasi-rectangular structure of the columnar phase. It is to be expected that the extent of interdigitation would be reduced at lower densities and this proves to be the case. At ρ^* of 2.5 the degree of interdigitation has decreased significantly, indeed to such an extent that it is now possible to pack six columns around a central column to give a hexagonal columnar phase. The nature of the positional ordering within a column was not investigated in any detail. However, the form of $g_\parallel(r_\parallel)$ for

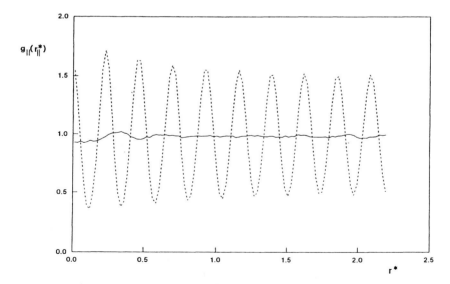

Fig. 11. The longitudinal distribution function $g_\parallel(r_\parallel)$ for the columnar phase at T^* of 2.5 (- - - - -) and the nematic phase at T^* of 4.1 (———)

the quasi-rectangular columnar phase (see Fig. 11) suggests that it is an ordered phase which would be consistent with the high degree of molecular interdigitation between neighbouring columns.

The unrealistic limiting form of the Gay-Berne potential employed by Emerson et al. [16] has prompted Bates and Luckhurst [17] to define a scaled and shifted distance

$$R = \left\{ r - \sigma(\hat{\mathbf{u}}_i \hat{\mathbf{u}}_j \hat{\mathbf{r}}) + \sigma_f \right\} / \sigma_f \tag{40}$$

for discotic systems which gives the potential a more appropriate limiting form at high energies. The potential energy curves obtained with this using the same values for κ, κ', μ and ν are shown in Fig. 10b. It is apparent that the well depths for particular molecular configurations are not changed. What is perhaps more obvious is that the widths of the attractive wells are considerably reduced so that the range of the attractive forces is significantly shorter. Associated with this change is a reduction in the separations corresponding to the minima in the potential for the various configurations. A system of Gay-Berne molecules interacting through this new form of the potential has been examined in considerable detail using an isothermal-isobaric ensemble. As we have seen in Sect. 5, simulations performed at constant pressure can have considerable advantages over their constant volume counterparts. Part of the phase diagram for the system was determined and revealed isotropic, nematic and columnar phases, as for the other Gay-Berne discotic [16]. The transition temperatures are found to increase linearly with increasing pressure, and the nematic range also increases, in accord with the Clapeyron equation.

The structure of the phases was studied in depth for the lowest scaled pressure of 25 for a larger system of 2000 molecules. It is, of course, of relevance to see whether the state points at which Gay-Berne systems are investigated correspond to those for real mesogens. As we saw in Sect. 5 it is possible to estimate the scaling parameters ε_0 and σ_0 used in simulations of Gay-Berne mesogens and using a similar procedure for the discotic system P^* of 25 is found to correspond to 55 atm, which is clearly reasonable although the nematic-isotropic transition temperature is somewhat high at about 470 °C. We begin with the nematic phase which is characterised by a radial distribution function devoid of structure at long range but with long range orientational order. This is clearly manifest for the singlet orientational distribution function, $f(\beta)$, which is strongly peaked when the molecular symmetry axes are parallel to the director. As we have found for rod-like systems the orientational distribution is well represented by the Maier-Saupe-like function given in Eq. (35). The distance dependent second rank orientational order parameter $\bar{P}_2^+(r^*)$ for the intermolecular vector is shown in Fig. 12 for the nematic phase. This is highly structured for short separations but tends to a limiting value of zero for r^* greater than about 3 indicating that the intermolecular vector is then distributed spherically. At short range the form of $\bar{P}_2^+(r^*)$ provides information about the local structure; thus at separations corresponding to the face-to-face arrangement $\bar{P}_2^+(r^*)$ is large and positive indicating that the intermolecular vector is highly ordered with respect to the director. In marked contrast at a separation of about 1, which is

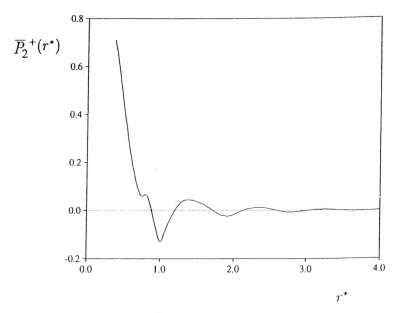

Fig. 12. The distance dependence of $\bar{P}_{2+}(r^*)$, the second rank orientational order parameter for the intermolecular vector in the nematic phase formed by GB(0.345, 0.2, 1, 2)at T^* of 2.6

associated with the edge-to-edge configuration, $\bar{P}_2^+(r^*)$ is negative which indicates that on average the intermolecular vector makes an angle greater than 54.74° with the director. However, it should be remembered that for this separation there may also be contributions from other molecular arrangements. This is certainly true for intermolecular separations greater than unity which necessarily complicates their interpretation.

The primary interest in this Gay-Berne discotic is the columnar structure, not only within the columnar phase itself but also in the preceding nematic and isotropic phases. This was explored using the columnar distribution function $g_c(r_{\parallel}^*)$ with the results shown in Fig. 13 for the three phases. In the isotropic phase there is a reasonably pronounced peak at $r_{\parallel}^*$ of 0.4 corresponding to the face-to-face configuration and stabilised by the deep well depth for this arrangement (see Fig. 10). Despite this strong attraction molecular columns are not formed in the isotropic phase which is shown by the absence of any other pronounced peaks. The local columnar structure is more obvious for the nematic phase where the orientational order makes it easier for molecules to pack into columns. The first peak is now clearly stronger and the second peak is apparent but beyond this the peaks are weak with vanishing intensity. In marked contrast the columnar distribution function in the columnar phase is highly structured with seven peaks being readily seen, confirming the assignment of the phase. To examine the longer range structure of the columnar phase it would be necessary to use a larger system size. The peak intensities decrease with increasing separation and the widths of the peaks increase, demonstrating a loss of correlation along the

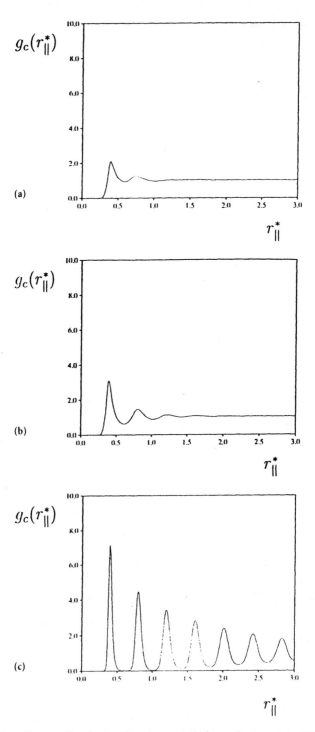

Fig. 13a–c. The columnar distribution function, $g_c(r_{\parallel}^*)$, for: **a** the isotropic ($T^* = 3.0$); **b** the nematic ($T^* = 2.6$); **c** the columnar ($T^* = 2.4$) phases of the Gay-Berne discotic GB(0.345, 0.2, 1, 2)

column. The intensities decay algebraically as $r_{\parallel}^{*-0.7}$ and not exponentially, in accord with the one dimensional character of the columns, although the system is not strictly one dimensional because of the weak interactions between molecules in neighbouring columns. The packing of the columns in the columnar phase was determined via the transverse distribution function $g_\perp(r_\perp^*)$. The peak at $r_\perp^*$ of zero coming from molecules in the same column is broad which indicates some positional disorder within the column. The next peak is at a separation of about one corresponding to the first shell of nearest neighbour columns; it is also broad showing that there is not quite perfect order. The next peak at $r_\perp^*$ of approximately 1.9 is broader than the first two and has a shoulder suggesting that it is in fact a combination of two peaks which is diagnostic of a hexagonal phase. Based on this combination of observations it seems reasonable to identify this as a hexagonal columnar phase with an ordering of molecules along the columns that is Col_{ho}.

Finally, we comment on the pronounced difference between the columnar phase formed by GB(0.345, 0.2, 1, 2) with a separation shifted and scaled with σ_e and that with σ_f. The Gay-Berne potentials based on the use of the contact distances σ_e and σ_f differ in two quite significant ways (see Fig. 10). Thus the range of the attractive part of the potential is considerably greater when σ_e is used rather than σ_f. Associated with this, the minimum in the potential occurs for larger intermolecular separations for the parametrisation using σ_e rather than σ_f. We attribute the strong interdigitation in the columnar phase found with σ_e to the greater separation of the potential minimum. As a consequence the average separation between molecules in a column is large, thus allowing a pronounced interdigitation by molecules in a neighbouring column.

7
Gay-Berne Molecules and Electrostatic Interactions

A common feature of many mesogenic molecules is the presence of polar substituents and aromatic cores [3]. The electrostatic interactions between such groups can be incorporated into a molecular potential with the addition of dipolar and quadrupolar terms, respectively. Rather than represent these permanent electrostatic interactions by using a model in which a charge distribution is scattered over the surface of the molecule, it is very common to use one (or more) point multipoles [2, 29]. Thus for an electrostatic Gay-Berne model, the pair potential is given by the sum

$$U(\hat{\mathbf{u}}_i\hat{\mathbf{u}}_j\mathbf{r}) = U_{GB}(\hat{\mathbf{u}}_i\hat{\mathbf{u}}_j\mathbf{r}) + U_{ELEC}(\hat{\mathbf{u}}_i\hat{\mathbf{u}}_j\mathbf{r}), \tag{41}$$

where $U_{ELEC}(\hat{\mathbf{u}}_i\hat{\mathbf{u}}_j\mathbf{r})$ is the multipolar expansion representing the electrostatic interaction. The interaction energy for a pair of point dipoles is [2, 29]

$$U_{DIP}(\hat{\mathbf{u}}_i\hat{\mathbf{u}}_j\mathbf{r}) = \frac{\mu_i\mu_j}{4\pi\varepsilon_o r^3}\left[(\hat{\mathbf{u}}_i.\hat{\mathbf{u}}_j) - 3(\hat{\mathbf{u}}_i.\hat{\mathbf{r}})(\hat{\mathbf{u}}_j.\hat{\mathbf{r}})\right], \tag{42}$$

in which μ_i is the dipole moment of molecule i and ε_0 is the permittivity of a vacuum which, given the context, should not be confused with the scaling parameter in the Gay-Berne potential. The corresponding expression for a pair of point quadrupoles is

$$U_{QUAD}(\hat{\mathbf{u}}_i\hat{\mathbf{u}}_j\mathbf{r}) = \frac{3Q_iQ_j}{16\pi\varepsilon_0 r^5} \left[1 + 2(\hat{\mathbf{u}}_i.\hat{\mathbf{u}}_j)^2 - 5(\hat{\mathbf{u}}_i.\hat{\mathbf{r}})^2 - 5(\hat{\mathbf{u}}_j.\hat{\mathbf{r}})^2 \right.$$
$$\left. -20(\hat{\mathbf{u}}_i.\hat{\mathbf{u}}_j)(\hat{\mathbf{u}}_i.\hat{\mathbf{r}})(\hat{\mathbf{u}}_j.\hat{\mathbf{r}}) + 35(\hat{\mathbf{u}}_i.\hat{\mathbf{r}})^2(\hat{\mathbf{u}}_j.\hat{\mathbf{r}})^2 \right],$$

(43)

in which Q_i is the uniaxial quadrupole moment of molecule i. Here we use the notation $\hat{\mathbf{u}}_i$ to represent the orientation of the multipole of molecule i, which need not coincide with the orientation of the symmetry axis of the molecule. Similarly, $\mathbf{r}$ is a vector of length r joining the two point multipoles, which may or may not be located at the centres of the molecules, and $\hat{\mathbf{r}}$ is a unit vector in this direction.

7.1
Dipolar Mesogens

7.1.1
Calamitic Systems

The influences of both longitudinal and transverse dipoles have been investigated for systems of calamitic molecules [61–64]. Satoh et al. [61, 62] have studied longitudinal point dipoles (i.e. $\mu_{i\,\parallel}$ $\hat{\mathbf{u}}_{ii}$) which are located either at the centre or shifted by σ_0 towards the end of the molecule for the GB(3.0, 5.0, 1, 2) model with a range of dipole moments. The simulations were performed using canonical Monte Carlo simulations for 256 molecules in a cubic box over a range of temperatures. For both types of model they report isotropic, nematic, smectic and crystalline phases although the smectic or crystalline phases were not characterised in any detail. Different types of behaviour for the nematic-isotropic transition as a function of dipole moment are reported for the two models. Thus, this transition is found to be shifted to higher temperatures for the terminal dipole model with increasing dipole moment, but not for the central dipole model. However, the smectic-nematic transition is shifted to higher temperatures for the central dipole model. They conclude that the central dipole has no influence on the orientational ordering process but does influence the stability of the smectic phase. This seems reasonable since the stability of the smectic phase will be enhanced by the tendency of the molecules to form nearest neighbour antiparallel side-by-side pairs, which are favoured by the dipolar interactions. However, at higher temperatures this arrangement does not appear to dominate the structure for the central dipole model and so the nematic phase is not stabilised with respect to the isotropic. Higher nematic-isotropic transition temperatures are observed for the terminal dipole model, in comparison to those found for the central dipole model, and these can be explained by the ellipsoidal shape of the molecule. Since the dipole is shifted towards the end of the tapered molecule, if two molecules are parallel or nearly parallel (which are expected to be dominant configurations in the nematic phase), then the shifted dipoles can approach closer than the central dipoles. In consequence the maximum possible well depth for a particular dipole moment is larger for the terminal dipole model and so the effective temperature at a particular scaled temperature will be lower; here we define an effective temperature in a similar way to the scaled temperature as $T_{eff} = k_B T/\varepsilon_{max}$, where ε_{max} is the strength of the largest interaction. For Gay-Berne models in which strong perturbations are added to the potential, this is an alternative way of scaling the temperature. For the conventional Gay-Berne potential, we recall that the scaled temperature is defined in terms of the Gay-Berne well depth, ε_o, for the cross configuration, which does not play a major role in the formation of the mesophases. However, the transition temperatures will depend strongly on the largest attractive interactions, since these interactions are, of course, the ones with high Boltzmann factors and so dominant in the determination of the structure of the phase. This feature of the model also appears to stabilise the nematic phase at higher temperatures as the dipole moment is increased. Presumably, increasing the strength of the dipolar interaction can induce antiparallel correlations and so leads to alignment in offset antiparallel pairs, which stabilises the nematic phase with respect to the isotropic at higher scaled temperatures as the effective temperature is reduced. Satoh et al. [61, 62] also claim that bilayers, with different layer spacings dependent on the dipole moment, are present in the smectic phase of the terminal dipole model. This structure would be formed by layers of molecules in which all of the dipoles in one layer point up and, in the next layer, the dipoles point down. Whilst this seems reasonable given the ability of the molecules to form offset pairs in an antiparallel arrangement, this observation becomes less clear cut when we consider the simulation conditions. The simulations were performed at constant volume in a box of fixed cubic shape for a system containing only 256 molecules. The small number of molecules implies that only a small number of layers were studied, for this system size typically only two or three complete layers. Indeed, the longitudinal pair correlation functions were only calculated to a maximum scaled separation of $r_{\parallel}^* \approx 4.0$, corresponding to just over one layer spacing which suggests that the box contains no more than three indistinguishable layers. The problem with this is what happens if the director is aligned along a box axis and there is an odd number of layers? The situation where cubic boxes are used and the director is not aligned along an arbitrary direction can lead to even worse problems for such small systems since this will almost certainly lead to incommensurate periodic images for the molecules in the bilayers. Thus the system is clearly not large enough to make confident predictions about the formation of bilayers, although the two different layer spacings for the neighbouring layers observed do hint at a bilayer structure. Just as importantly, the box shape is not allowed to change during the simulation which can lead to inappropriate deformations of the smectic structure, a feature of liquid crystal simulations which has now been known for many years, especially for such small system sizes [10].

Similar methodological constraints are likely to affect the results of the simulations of molecules with transverse dipoles by Gwózdz et al. [63]. These molecular dynamics simulations were performed along an isotherm by gradually compressing the system. It is claimed that such systems exhibit a tilted smectic phase, in which the tilt angle varies between approximately 2 and 10°; the exact tilt angle is found to be strongly dependent on density. However, they also observe a similar behaviour in the smectic phase of a non-polar system, although the exact values and the temperature dependence of the tilt angles are slightly different. We note that this behaviour was also observed by de Miguel et al. [44] in one of the first simulations of a Gay-Berne model, but is now believed to be due to the use of short equilibration runs and the constraint of a cubic box on a small system. The observation of this phenomenon by Gwózdz et al. [63] led to the claim that the transverse dipolar model exhibits a tilted smectic phase, with a structure intermediate between the smectic I and smectic F. However, snapshots from the simulation clearly indicate that the positions of the molecules do not change either during the individual simulations or from one simulation to the next at a higher density. The fact that the molecules hardly diffuse during the simulations at a single state point is, of course, entirely consistent with the extremely low diffusion coefficients exhibited by real systems in the highly ordered smectic phases. Nonetheless, longer simulations should be performed to ensure equilibration. The fact that the relative positions do not differ after compression of the system is much more serious. Since the rotational motion of both the molecules and the director are essentially frozen at the high densities characteristic of the smectic B phase, the layers cannot rotate in the box. This fact, combined with the constant cubic box shape, means that as the box is compressed from one run to the next, a reduction of the layer spacing must occur. As the density is further increased this will eventually lead to an overlap between the layers. An obvious question then is what do the molecules do to avoid the high potential energies associated with this overlap? The only reasonable explanation is that the molecules tilt to reduce the effective layer spacing.

Berardi et al. [64] have investigated the structure of the smectic phases exhibited by molecules with longitudinal dipoles for the GB(3.0, 5.0, 1, 3) mesogen. They have studied three state points for both central and shifted dipoles for a single dipole moment using canonical Monte Carlo simulations on a system of 1000 molecules in a cubic box. Of course, the use of larger systems and longer equilibration runs, of up to 200,000 cycles, means that much more computer time will be spent on each state point and so fewer state points can be studied. The simulation of the central dipole model gives a smectic structure very similar to that observed by Satoh et al. [61]; that is, within the smectic layer, the molecules tend to align on average with antiparallel side-by-side neighbours. The second peak in the radial distribution function for the smectic phase is found to be split, which is characteristic of in-plane hexagonal order, and so the smectic phase was characterised as a smectic B. They also found that the longitudinal translational distribution function is narrower for the central dipolar system than for a non-polar system at the same reduced temperature and that the layer spacing is larger in the polar system. The narrower distribution is presumably due to the stronger side-by-side interaction in the dipolar model which leads to a lower effective temperature. However, as we noted earlier, the concept of reduced temperature is rather imprecise, since the temperature is scaled by the well depth of the Gay-Berne cross configuration and so does not take the possibly strong dipolar interaction into account. The reason for the difference in the layer spacings is not so clear, since the simulations were performed in cubic boxes of fixed dimensions, for the same number of molecules at the same density, and so this should not vary unless cavities are formed. Indeed, the fact that the molecules are quite closely packed within the smectic B layers means that, at the same density, the layer spacing should be larger than in a less well-ordered smectic A phase, since the in-plane density would be larger and so there would be fewer layers for a given number of molecules; however, possible reasons for this change in layer spacing were not considered. The simulation for the shifted dipole system indicates that a very different smectic structure is formed for terminal dipoles. Thus, molecules in a single layer tend to align parallel. The net polarisation caused by this alignment is compensated for by the molecules in adjacent layers aligning in the opposite direction. This leads to a bilayer structure in which the molecules pack with two different repeat distances; the head-to-head repeat distance is less than the tail-to-tail. Since parallel side-by-side neighbours are likely to be destabilised by the dipolar interaction, we may wonder why a smectic phase is observed for this model. Indeed, it seems reasonable that the strong Boltzmann factor which leads to enhanced correlations between offset antiparallel molecules is likely to be the driving force for the formation of the smectic phase, especially if the head-to-head layers are interdigitated, since this increases the distance between parallel side-by-side dipoles. This structure is clearly similar to that observed by Satoh et al. [62] for terminal dipoles. However, Berardi et al. [64] find that the smectic phase is not truly a bilayer structure, but that the system has a stripe-like domain structure which is shown in Fig. 14. The dimensions of the stripes are larger than the box size used by Satoh et al. [62] which explains why such behaviour was not observed in the smaller system; this emphasises our earlier comment that relatively large systems should be used, at least at a few state points, to check the behaviour found for small systems. This rather striking structure was also found to be stable in an even larger system of 8000 molecules. It also appears to be more stable than a smectic in which alternating layers have all the dipoles pointing in the same direction (i.e. no stripes), since a simulation started from this latter structure was found to transform into the striped structure. Of course, using conventional Monte Carlo moves, this would never occur because the rotational motion is extremely hindered in such a dense phase. This was overcome by allowing non-physical moves, involving flipping the orientation of a molecule, which leads to a more efficient equilibration scheme than that using purely physical moves. In a Monte Carlo simulation, these moves are, of course, perfectly valid since the time evolution

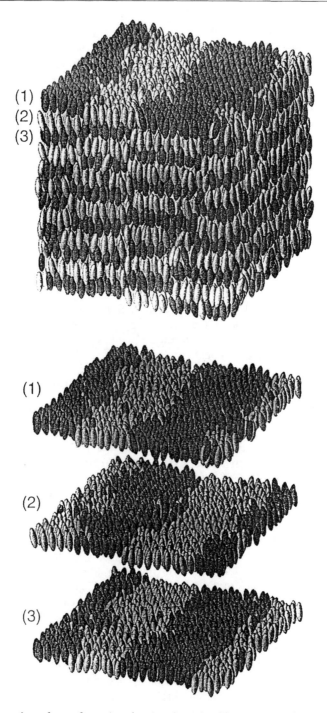

Fig. 14. A snapshot of a configuration showing the stripe-like structure of the smectic phase formed by the polar mesogen GB(3.0, 5.0, 1, 3) and the antiferroelectric compensation in adjacent layers. The different orientations of the dipoles are indicated by the different shading of the ellipsoids

of the system is not followed. However, there still appear to be unanswered questions about this structure. For example, the boundaries between the stripes appear to align parallel to the box edges and to each other. This presumably occurs because the orientations of the stripes must be commensurate with the boundary conditions. It would be interesting to discover if the stripes are always aligned parallel to each other for larger systems since, if not, this should lead to interesting defects where the boundaries meet.

The problems highlighted here clearly indicate that, to draw any conclusions about important features such as bilayer formation and tilted structures in the smectic phases of dipolar models, large systems should be used so that distribution functions can be calculated for separations greater than one layer spacing and, most importantly, the layer spacing of the smectic phases should be allowed to fluctuate so that it can achieve its natural value. This requirement can be attained most easily in constant pressure simulations. The use of constant pressure simulations also rules out the possibility that the system is at a density which is within the coexistence density range of two phases [14], a feature which is often overlooked by many authors and so leads to the values of observables such as the second rank orientational order parameter at the nematic-isotropic transition being underestimated. This is because the system may be kept at a density at which it would naturally separate into two bulk phases of different density, but it is prevented from doing so by finite size effects [65] and so the order parameter increases almost continuously through the coexistence region. For the smectic phases, alignment of the layer normal along a box axis in a constant pressure simulation also allows the layer spacing to fluctuate independently of the in-plane separation and also ensures that the smectic layers are perfectly commensurate with each other [43]. Indeed, this method was used in simulations of the GB(4.4, 20.0, 1, 1) mesogen [14] which revealed that the layer spacing for this model does not depend on temperature and so is also essentially independent of the density; this feature would not have been apparent if the system was studied under the conditions of constant box shape and the director orientation had been essentially frozen (within the simulation time) along some arbitrary direction, since the layer spacing under these conditions would be highly density dependent. This convenient alignment can be easily attained by applying an aligning field at a temperature just above the transition to the smectic phase and then cooling below the transition. Of course, the transition temperature must have been determined in previous simulations without the field and extended equilibration runs must be performed after the field is turned off to remove any extra orientational order induced by the field.

We conclude that the dipolar models may exhibit many interesting features, such as the influence of the strength of the dipole moment on the nematic-isotropic and smectic-nematic transitions. Determining the structure of the various smectic phases is clearly an area which needs more investigation. However, an appropriate simulation scheme must be used to avoid misunderstanding the behaviour caused by the method rather than the model. This may be as simple as checking for system size dependence, or running simulations at the same state points starting from different initial configurations.

We note that we have not mentioned how the long range nature of the dipolar interaction is handled in these simulations. There appears to be three methods applied in the literature. Thus Berardi et al. [64] use the Ewald-sum technique [2], which tends to over-emphasise slightly the periodic nature of the fluid. Satoh et al. [61, 62] choose the reaction field method [2], which assumes an average electrostatic interaction beyond a certain cut-off distance and so requires an a priori estimate of the relative permittivity of the fluid. However, a more approximate method is used by Gwózdz et al. [63] in which the pair potential is simply cut at half the box length. For short ranged potentials or very large systems, this should not be a problem but, for this model and small system size, the long range nature of the dipolar interaction is completely neglected. This practice also leads to a slightly different potential being studied at each density; thus for small system sizes of 256 molecules, the potential is cut and shifted at $r_c^* \approx 4.08$ at the highest density studied ($\rho^* = 0.47$) but at $r_c^* \approx 6.84$ for the lowest density ($\rho^* = 0.10$). A better approach would be to determine the potential cut-and-shift distance and ensure that the box is large enough at each density. However, we may wonder what influence the long range nature of the dipolar interaction has on the structure of the phases; that is, do the structural features in these systems arise from local correlations or are long range correlations also important? Clearly more work is necessary to answer this intriguing question. It may be that the use of the spherical cut-off at short distances of the order of just over a molecular length may be a legitimate route to determine the approximate phase behaviour of a model, but that the more sophisticated methods of including the long range tail of the dipolar potential must be used at selected, interesting state points to check the validity of the simulations. This may be especially important for the stripe textures observed by Berardi et al. [64] which appear to have fairly large dimensions in comparison to the molecular size. The influence of the method of computing the long range part of the potential on the structure also requires further investigation; thus it is not known if the stripes also occur using the reaction field method, or indeed if they occur if the potential is restricted to short range contributions.

7.1.2
Discotic Systems

Berardi et al. [66] have also investigated the influence of central dipoles in discotic molecules. This system was studied using canonical Monte Carlo simulations at constant density over a range of temperatures for a system of 1000 molecules. Just as in discotic systems with no dipolar interaction, isotropic, nematic and columnar phases are observed, although at the low density studied the columnar phase has cavities within the structure. This effect was discovered in an earlier constant density investigation of the phase behaviour of discotic Gay-Berne molecules and is due to the significant difference between the natural densities of the columnar and nematic phases

[17]. Thus the density change at the transition between these phases for non-polar models at constant pressure is typically more than 10% [17]. Accordingly, if a low density is selected for canonical ensemble simulations, then the columnar phase will contain cavities. In contrast, if the density is sufficiently high to remove the cavities then the columnar phase will melt only at extremely high temperatures. Thus the constant pressure ensemble is more appropriate for model systems where the natural density of the various phases may differ significantly. However, the formation of the cavities noted for dipolar Gay-Berne discs [66] was simply ignored and no attempt was made to study other densities where cavities were not present. The structures of the nematic and isotropic phases for the dipole model appear to be essentially unchanged from their non-polar counterparts and so the dipolar interaction does not appear to affect the translationally disordered phases. At low temperatures, the columnar phase is found to consist of columns of molecules in which the dipoles tend to be aligned, on average, parallel. However, these polar domains are not found to extend over the full length of the columns. Also neighbouring columns are observed to have a random up-down alignment and so the system does not exhibit a ferroelectric phase. However, for very low temperatures, the presence of the domains leads to an imbalance of up and down columns and so to a moderately ferroelectric columnar phase, but this is likely to be an artefact of the system size. As we have noted, larger systems would be needed to clarify this point.

7.2
Quadrupolar Mesogens

7.2.1
Calamitic Systems

A number of models have been proposed to explain the formation of layered phases in which the director is tilted away from the layer normal such as the smectic C. The majority of the explanations rely on the fact that the model molecules must possess biaxial symmetry in keeping with the symmetry of the phase. One model consists of a pair of dipoles pointing in opposite directions, inclined with respect to the long axis of the molecule [67]. This model seems entirely reasonable; indeed many mesogens have large off-axis dipoles. However, other compounds which do not possess such dipoles also exhibit smectic C phases, although there is clearly no reason why the same mechanism is responsible for the tilt in all smectic C phases. A further molecular feature suggested to result in the formation of the smectic C is the presence of two terminal alkyl chains, which give the molecule a zigzag shape. This geometrical model seems reasonable since many mesogens possess such features. An additional explanation for the creation of tilted phases has been suggested, which is quite intriguing since it is based on axially symmetric molecules [68]. Thus a smectic C phase is predicted for uniaxial molecules which interact through an electrostatic quadrupolar potential.

The effect of a point quadrupole on the behaviour of the GB(4.4, 20.0, 1, 1) mesogen has been investigated by Bacchiocchi et al. [69]. The quadrupole moment is taken to have uniaxial symmetry and to be aligned along the molecular symmetry axis and positioned at the centre of the molecule, which preserves the $D_{\infty h}$ symmetry of the original Gay-Berne potential. This model was studied at a pressure for which the bare GB(4.4, 20.0, 1, 1) model exhibits smectic B, smectic A and isotropic phases [14]; presumably a crystal phase is also present at temperatures below the smectic B, but the high densities at which this transition is expected to take place means that this has not been studied. The quadrupolar system was found to exhibit a nematic phase on cooling from the isotropic. The reduced temperature at which this transition was found to occur is similar to that observed for the smectic A-isotropic transition for the bare Gay-Berne model. However, even though the translational order for the smectic A phase of the GB(4.4, 20.0, 1, 1) mesogen is destroyed by the quadrupolar interaction, the orientational order does not appear to be affected and so a nematic is observed in place of the smectic A phase. On further cooling, the system is found to exhibit a tilted smectic phase. Since there is no in-plane order within the smectic layers, this phase is characterised as a smectic C. No evidence for a smectic A phase was found. This indicates that at sufficiently high but not too high temperatures, the quadrupolar interaction, which tends to destabilise the side-to-side arrangement, disrupts the positional order present in the smectic A phase of the GB(4.4, 20.0, 1, 1) mesogen and so leads to the formation of the nematic phase. However, the presence of potential minima for shifted parallel configurations favours the formation of a tilted, layered structure at lower temperatures. Even though the potential is axially symmetric, a single tilt direction appears to be favoured by the system. This single tilt direction is observed to be adopted by each layer; that is, the directors in different layers tilt in the same direction; thus the director is fixed and the layers tilt. This behaviour is to be expected because there is a nematic phase above the smectic C and so the molecules tend to align in the same direction before the layered structure is formed. It may also be possible that such an arrangement could lead to a smectic C phase in which a chevron structure is formed [3]. However, this structure must fit into the periodic boundaries like any other translationally ordered system and so will probably not be observed for the relatively small systems available for simulation which in this case used 2000 molecules. If the nematic phase was not present, it may be that the layers would form with different tilt directions of the director in each layer. At lower temperatures still, the system exhibits a tilted analogue of the smectic B phase. This is characterised as either a smectic I or a smectic F [3]; the distribution functions were not able, however, to distinguish between the subtle differences of these two phases.

7.2.2
Discotic Systems

A similar model has been used to study the effect of a quadrupolar interaction on a system of Gay-Berne discs [27]. In this case, the bare Gay-Berne model was found to exhibit nematic and hexagonal columnar mesophases as we saw in Sect. 6. The addition of a weak quadrupole moment was found not to modify the phase behaviour; here the strength of the quadrupolar interaction was about 5% of that of the side-to-side Gay-Berne interaction. However, doubling the quadrupole moment (leading to a quadrupolar interaction of about 40% of the side-to-side interaction) was found to have a profound influence; thus the columnar phase is destabilised with respect to the nematic. Just as the quadrupole leads to a destabilisation of the interaction between a pair of side-by-side rods, the equivalent term in the discotic potential reduces the face-to-face interaction between discs. This leads to the potential minimum for a pair of discs being shifted from the face-to-face position to an off-axis position; the potential minimum for parallel molecules is shifted by about 30° [27]. It might be expected that the quadrupolar interaction could lead to a tilted columnar phase, which would be the discotic equivalent of the tilted smectic C phase observed for quadrupolar rods. However, such a phase was not observed.

The primary reason for studying quadrupolar discs was to examine the possibility of phase induction in mixtures of such molecules. New liquid crystal phases frequently occur in binary mixtures of materials. This usually results from the depression of the melting point and so to the observation of phases which would otherwise be obscured by the system freezing. However, more intriguing is the observation of chemically-induced phases which occur above the melting point of either component. One of the first observations of a chemically-induced liquid crystal phase was for mixtures of multiynes and 2,4,7-trinitrofluorenone [70]. Neither of these components is mesogenic, but their binary mixtures exhibit nematic and columnar phases, depending on the exact composition. Similarly, an equimolar mixture of benzene and hexafluorobenzene, which are both liquids at room temperature, exhibits a chemically-induced crystalline phase [71] in which the structure is essentially columnar with alternating benzene and hexafluorobenzene molecules. This observation is consistent with the fact that the quadrupolar interaction plays a dominant role in the organisation of such molecules. To explore the idea that a quadrupolar interaction may be responsible for the induced liquid crystalline phases observed by Praefcke et al. [70], simulations have been performed on mixtures of quadrupolar discs, in which the two components differ only in the sign of the quadrupole moment. The results of the simulations of the mixtures were in marked contrast to those for the pure systems. For an equimolar mixture of quadrupolar discs, the system forms a columnar phase directly from the isotropic, at a temperature which is higher than the nematic-isotropic transition of the pure system. A strong tendency for unlike particles to be nearest neighbours within the columns is observed, giving a structure in which the particles alternate along the columns. However, the phase behaviour is even more interesting for an unsymmetric mixture. Thus a 75:25 mixture exhibits a nematic phase between the columnar and isotropic phase. This nematic phase is particularly intriguing since it has a different structure to that observed for the nematic of the pure system. In the mixture, the nematic appears to be formed from small columns of approximately three or four molecules with the different components of the mixture alternating along the column axis. The short columns align in the same way as calamitic molecules to form the nematic phase, as in the original structure proposed by Praefcke et al. [70]. This local structure contrasts with the extremely short range translational order usually observed in nematic phases of discotic molecules. Thus this simulation indicates that the quadrupolar interaction between the different large aromatic fragments gives a possible mechanism for the induction of liquid crystal systems in mixtures of disc-shaped molecules and that it is not necessary to invoke a charge-transfer interaction.

8
Chiral Gay-Berne Mesogens

The structures of phases such as the chiral nematic, the blue phases and the twist grain boundary phases are known to result from the presence of chiral interactions between the constituent molecules [3]. It should be possible, therefore, to explore the properties of such phases with computer simulations by introducing chirality into the pair potential and this can be achieved in two quite different ways. In one a point chiral interaction is added to the Gay-Berne potential in essentially the same manner as electrostatic interactions have been included (see Sect. 7). In the other, quite different approach a chiral molecule is created by linking together two or more Gay-Berne particles as in the formation of biaxial molecules (see Sect. 10). Here we shall consider the phases formed by chiral Gay-Berne systems produced using both strategies.

The simplest form for the chiral interaction between two cylindrically symmetric particles is [72]

$$U_c(\hat{\mathbf{u}}_i\hat{\mathbf{u}}_j\mathbf{r}) = K(r)(\hat{\mathbf{u}}_i.\hat{\mathbf{u}}_j)(\hat{\mathbf{u}}_i \times \hat{\mathbf{u}}_j.\hat{\mathbf{r}}), \tag{44}$$

where the strength of the interaction depends on the separation between the point centres and the molecular chirality. A variety of models have been developed to relate $K(r)$ to these factors [73] and one based on a chiral distribution of sites with linear polarizabilities shows that $K(r)$ varies as r^{-7}. This has prompted Memmer et al. [74] to write the chiral interaction between two Gay-Berne particles as

$$U_c(\hat{\mathbf{u}}_i\hat{\mathbf{u}}_j\mathbf{r}) = 4\varepsilon(\hat{\mathbf{u}}_i\hat{\mathbf{u}}_j\hat{\mathbf{r}})R^{-7}(\hat{\mathbf{u}}_i.\hat{\mathbf{u}}_j)(\hat{\mathbf{u}}_i \times \hat{\mathbf{u}}_j.\hat{\mathbf{r}}), \tag{45}$$

where the scaled and shifted separation, R, is given by Eq. (5). The orientational dependence of both R and $\varepsilon(\hat{\mathbf{u}}_i\hat{\mathbf{u}}_j\hat{\mathbf{r}})$ means that this chiral interaction cannot be thought of as originating from point centres, although Tsykalo [75] has used a true point interaction albeit imbedded in the Berne-Pechukas potential [11]. To see the significance of the chiral interaction it is helpful to consider the form of the interaction for two molecules when they are orthogonal to the intermolecular vector. The potential is a minimum when the angle between the molecular symmetry axes is about 45°. The location of the minimum which controls to a large extent the nature of the twisted structures can be shifted by the addition of an achiral potential represented by the Gay-Berne potential for which the energy is a minimum when the molecules are parallel for the same configuration. The total potential is then

$$U(\hat{\mathbf{u}}_i\hat{\mathbf{u}}_j\mathbf{r}) = U_a(\hat{\mathbf{u}}_i\hat{\mathbf{u}}_j\mathbf{r}) + cU_c(\hat{\mathbf{u}}_i\hat{\mathbf{u}}_j\mathbf{r}), \tag{46}$$

where the parameter c controls the relative strength of the chiral interaction.

The parametrisation chosen for the Gay-Berne interaction was that proposed by Luckhurst et al. [10], namely GB(3.0, 5.0, 1, 2) for which the phase behaviour is relatively well-understood, as we have seen in Sect. 5. At a density ρ^* of 0.30 and a temperature T^* of 1.50 the achiral system forms a highly ordered nematic phase with $\overline{P}_2$ equal to about 0.8. To explore the range of values of the chirality parameter likely to lead to chiral phases, this state point was studied as a function of c. It was found that for c less than about 0.65 the nematic phase is stable but above this value a chiral nematic with its characteristic helical distribution of the director is formed. This structure was identified by visualising the configurations taken from the end of the production run with the results shown in Fig. 15a viewed along the three edges of the simulation box. The molecules are drawn as lines projected onto the plane orthogonal to the viewing axis. In this way the helical structure of the phase is readily apparent although it is clear that the inclusion of the neighbouring periodic images of the simulations assists in the creation of an image with long range order. The helix axis is seen to be aligned along a box edge and the pitch of the helix is essentially twice the box dimension. Both observations are somewhat surprising and are thought to be a consequence of the use of cubic periodic boundary conditions [74]. The use of these means that because of the spatially varying structure the periodic images will not necessarily be commensurate with the molecules in the simulation box, which increases the energy of the system. To resolve this problem the chiral nematic will not only adjust the pitch of the helix but also the orientation of the helix axis; such behaviour being analogous to that encountered for smectic phases in Sect. 5. This explanation is also consistent with the observation that the system changes sharply from a nematic to a chiral nematic as c increases; presumably the pitch of the helix expected for smaller values of c is too large to be accommodated in the simulation box but the gain in energy in decreasing the pitch is too great. It would seem, therefore, that the helical structure of the chiral nematic formed by the chiral GB(3.0, 5.0, 1, 2) is determined to a significant extent by the periodic boundary conditions and so does not reflect the natural structure of the phase. This difficulty can be overcome by using different periodic boundary conditions to take account of the helical structure [76]; indeed some of these have been found to work successfully for lattice models of chiral nematics [77].

Increasing the chirality parameter produces another sharp change in the structure of the system when c is about 1; a snapshot showing the molecular organisation in the new phase is given in Fig. 15b. It is apparent that the images obtained by viewing the configuration along the three orthogonal box edges are identical and apparently devoid of long range orientational order. It

(a)   *(b)*

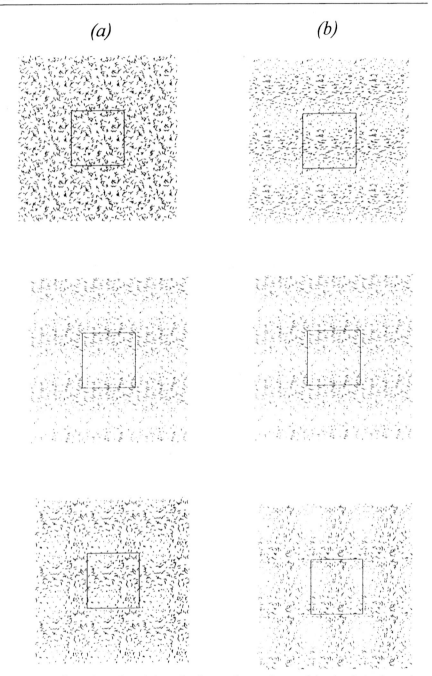

Fig. 15a,b. Configurations viewed along the three orthogonal axes of the simulation box taken from the production run for chiral GB(3.0, 5.0, 1, 2) at ρ^* of 0.30 and T^* of 1.50 with the chiral parameter c set equal to: **a** 0.9; **b** 1.3. The neighbouring periodic images of the simulation box are also shown

(a)   (b)   (c)

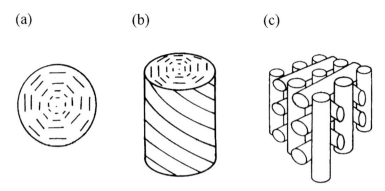

Fig. 16a-c. The model proposed for blue phase II showing: **a,b** the director arrangement within a double twist cylinder; **c** the packing of these cylinders along three orthogonal axes

would seem that the increase in the molecular chirality has resulted in the formation of a blue phase [3] which is to be expected because, experimentally, it is found that there is a minimum helical pitch necessary if a chiral nematic is to undergo a transition to a blue phase [78] and the pitch is inversely proportional to the molecular chirality. There are three blue phases and the structure of blue phase II, sketched in Fig. 16, is composed of double twist cylinders formed by twisting the director around the cylinder. These double twist cylinders are then packed along three orthogonal axes to give the structure of blue phase II. The molecular arrangement in such a structure is clearly reminiscent of that found for chiral GB(3.0, 5.0, 1, 2) and is shown in Fig. 15b.

Once the range of the chirality parameter c necessary to yield chiral phases had been established it was then possible to see how the phase structure changes with temperature, as for real systems. The parameter c was set equal to 0.8 and the phase behaviour studied. The system is found to form a chiral nematic phase as anticipated but we also expect to observe a transition to a smectic phase since the achiral Gay-Berne mesogen does form a smectic B phase (see Sect. 5). A smectic phase is certainly formed at lower temperatures and the molecular organisation in this phase is shown in Fig. 17. It is suggested [74] that this phase is a twist grain boundary phase in which blocks of smectic phase are arranged with their layer normals tracing out a helical path [3]. However, to make this possible the gaps separating the smectic blocks consist of defects which remove the layer structure. The molecular organisation shown in Fig. 17 is claimed to be consistent with the structure of the twist grain boundary phase. This is, in principle, a fascinating result although it is clear that to establish the structure in sufficient detail will require a considerably larger system than the 256 molecules employed in these simulations [74]. In addition an NPT rather than an NVT ensemble should be used together with spiralling boundary conditions to ensure that the system adopts the natural structure of the phase.

We now consider the alternative strategy with which to create chiral molecules using Gay-Berne molecules as a basis. In this two rod-like molecules are linked together resulting in a wide range of chiral conformations. One of

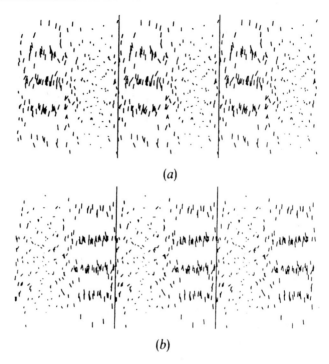

Fig. 17a,b. Snapshots viewed along orthogonal directions showing the molecular organisation within the smectic phase formed by chiral GB(3.0, 5.0, 1, 2) with c equal to 0.8 at ρ^* of 0.30 and T^* of 0.75; the neighbouring periodic images are also included

the simplest has the symmetry axes of the two Gay-Berne molecules orthogonal to the intermolecular vector, making a fixed torsional angle φ and with the separation between the molecules constant [79]. Provided φ is neither $0°$ or $90°$ the composite molecule is chiral with $\varphi < 0°$ having M-helicity. Real molecules with such twisted structures when used as dopants are known to convert nematic to chiral nematic phases and to have very high helical twisting powers [80]. Compounds composed of these twisted molecules do not form liquid crystal phases themselves although, as we shall see, this is not the case for molecules formed by linking two Gay-Berne particles together.

In the NVT Monte Carlo simulations of such systems two GB(3.0, 5.0, 1, 2) molecules were linked together with a torsional angle of $±30°$ and a scaled separation between the two of 1.1 which is approximately the width of a Gay-Berne molecule. The system also contained a small number ($\sim7\%$) of guest molecules again formed by linking two Gay-Berne molecules together but whose intramolecular energy was independent of the torsional angle. Their presence was to investigate the transfer of chirality from the phase to the guest, as we shall see shortly. The density and temperature used in the simulation were comparable to those at which the Gay-Berne mesogen GB(3.0, 5.0, 1, 2) is found to exhibit a nematic phase [10]. As for the system with a point chiral centre, standard periodic boundary conditions were used, which may well influence the detailed phase behaviour. In any event the system is found to form a chiral nematic phase as had been anticipated. The pitch of the helical structure was studied using a longitudinal orientational correlation coefficient which is analogous to that introduced in Sect. 3 (see Eq. (19)) and is defined by [74]

$$G_2(r_\parallel) = \overline{P_2(\cos \beta_{ij})} . \tag{47}$$

Here, $r_\parallel$ is the separation between the molecules resolved along the helix axis and β_{ij} is the angle between an appropriate molecular axis in the two chiral molecules. For this system the $C_2^{(3)}$ axis closest to the symmetry axes of the constituent Gay-Berne molecules is used. In the chiral nematic phase $G_2(r_\parallel)$ is periodic with a periodicity equal to half the pitch of the helix. For this system, like that with a point chiral centre, the pitch of the helix is approximately twice the dimensions of the simulation box. This clearly shows the influence of the periodic boundary conditions on the structure of the phase formed [74]. As we would expect simulations using the atropisomer with the opposite helicity simply reverses the sense of the helix.

   Although the formation of a chiral nematic phase is of considerable interest, the main point of the study was to explore the transfer of chirality from the host to the guest. It is such transfer, albeit in reverse, which is thought to be responsible for the large helical twisting powers of chiral biaryls [80]; it is argued that the chiral guest stabilises chiral conformations of the host molecules. In the simulation study it is found that the chiral nematic phase does produce a chirality in the guest by favouring chiral conformers. That is, the distribution function for the torsional angle is no longer independent of this angle but is peaked at a value of about 10° for the chiral host with P-helicity and, necessarily, at −10° for the host with the opposite, M-helicity.

9
Flexible Gay-Berne Molecules

A feature of real liquid crystals which is omitted from the Gay-Berne potential is flexibility. Whilst some mesogens, such as 4,4'-dimethoxyazoxbenzene, are essentially rigid or at least have conformers with similar shapes, others such as the 4-alkyl-4'-cyanobiphenyls have a flexible alkyl tail. Such molecular features are found to play a key role in increasing the nematic range by lowering the melting point and in stabilising the smectic phase by increasing the smectic-nematic or smectic-isotropic transition temperature [3]. A number of recent simulation studies have attempted to take into account such features by the use of potentials with atomic detail [6]. These models are typically based on a molecular skeleton and so represent a particular compound in a fairly precise way. The hydrogen atoms are not usually included but united into the atom to which they are attached, since these are small (and so are not thought to be important for determining the phase structure) but they are numerous in typical mesogenic molecules (and so expensive to include). However, even within the united atom approach, these models have a high computational cost and so only a small number of molecules, of the order of 100, can usually be studied. This has quite serious implications for the investigation of liquid crystals, since it means that correlations over only small distances can be studied. In addition, to simulate the internal molecular structure accurately, a small time step must be used, typically 2 fs, which is an order of magnitude smaller than those used for rigid body motion for single site models. As a consequence, atomistic detailed models can be used to equilibrate and study the internal structure of a mesogen at a few state points, but not to explore the phase behaviour. Indeed, to date only one study of an atomistic detailed model has shown the spontaneous formation of a nematic phase from an isotropic phase [81].

However, it is possible to borrow the ideas of an atom-based potential and to include these in a more approximate framework. Thus the effects of alkyl chains can be accommodated in mesogenic models using the Gay-Berne potential; the rigid core of the mesogen can be modelled by a Gay-Berne unit, with a linear array of Lennard-Jones sites attached to represent the alkyl fragment. This has been done by La Penna et al. [82] for a series of generic liquid crystals composed of a single rigid core and a single terminal alkyl chain; the different members of the series have chain lengths of four, six or eight combined atoms. Wilson [83] has also used a similar model of two Gay-Berne units linked by spacer composed of Lennard-Jones sites to study liquid crystal dimers where the dependence of the transitional properties on the parity of the number of atoms in the spacer is especially pronounced [84]. The interaction energy between a pair of such molecules is calculated in a pseudo-atomistic way. Thus the calculation involves the summation over all non-bonded GB-GB, LJ-LJ and GB-LJ units and, within each molecule, over all harmonic 1–3 bending potentials and 1–4 torsional potentials (see, for example, [85]). The total non-bonded potential energy U_{nb} is

$$U_{nb} = \sum_{(i,j)} U_{LJ-LJ} + \sum_{(i,j)} U_{GB-GB} + \sum_{(i,j)} U_{GB-LJ}, \tag{48}$$

in which the notation $\sum_{(i,j)}$ implies the summation over all non-bonded pairs (i,j). The bond angle bending $U_{1,3}$ and torsional $U_{1,4}$ potentials, which determine the flexibility, have the form

$$U_{1,3} + U_{1,4} = \sum_{(i,j,k)} k_{\theta}(\theta - \theta_o)^2 + \sum_{(i,j,k,l)} k_{\phi}^n (1 + \cos(n\phi - \phi_o)), \tag{49}$$

or similar equivalent forms, in which $\sum_{(i,j,k)}$ and $\sum_{(i,j,k,l)}$ imply summations over all bending and torsional angles respectively, k_{θ} and k_{ϕ}^n are force constants for the bending and torsional potentials and θ_o and ϕ_o are the equilibrium angles. The simulations of La Penna et al. [82] and Wilson [83] differ in the way they calculate the interaction between a Gay-Berne unit and a Lennard-Jones site. For this calculation, La Penna et al. [82], following Emsley et al. [86], replace the Gay-Berne unit with a linear array of four Lennard-Jones sites, as used originally by Gay and Berne to parametrise their potential; the usual Gay-Berne potential is used for the GB-GB interactions. However, this is clearly not strictly appropriate. The approach adopted by Wilson is more consistent and the generalised Gay-Berne potential is used to model the interaction between different types of site.

The addition of the chain clearly introduces more realism into the generic Gay-Berne potential for mesogenic molecules. The results from such simulations can, therefore, be tentatively compared to those for specific mesogens. Thus, for the nematic phase exhibited by their model, La Penna et al. [82] have calculated segmental second rank orientational order parameters for various bonds along the chain. These are found to exhibit a strong odd-even effect as we can see from the results in Fig. 18. Such alternations compare well with those determined from NMR experiments. Of course, since much of the atomic detail is lost from the model, the results cannot be simply compared to a particular mesogen and expected to give good agreement. However, such a model should give a general idea of how the alkyl chain affects, for example, the orientational order in the nematic phase and in turn how the conformational order influences this. The flexible model can also be used to study the conformation of the chain in the nematic phase, as done by La Penna et al. [82]. The fact that the results obtained are similar to those from experimental studies clearly indicates the success of such simple models. Indeed, many of the fine atomic details appear not to be necessary and a generic potential can reproduce many (but presumably not all) features of liquid crystal molecules.

The dimer model proposed by Wilson [83] was found not to exhibit a nematic phase but smectic A and B phases and so, although semi-realistic features are included, the model is not able to reproduce the behaviour of real dimers such as the α,ω-bis(4'-cyanobiphenyl-4-yloxy) [87], which exhibit nematic phases, although liquid crystal dimers exhibiting smectic phases are also known [88]. In these simulations, the GB-GB interaction appears to be too strong and so the system forms a smectic phase in which the GB and LJ units

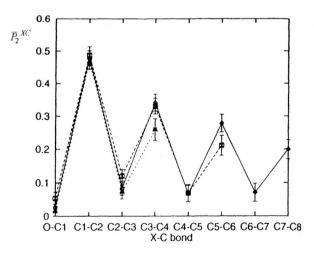

Fig. 18. The dependence of the second rank orientational order parameters, $\bar{P}_2^{XC}$, for the X-C bond on its position along the alkyl chain containing four ($\triangle$), six ($\square$) and eight ($\blacklozenge$) atoms

undergo microphase segregation. Wilson notes that it is possible to increase the strength of the GB-LJ interaction to prevent the segregation and possibly lead to the formation of a nematic phase. However, it seems that a better approach is to follow the method of Luckhurst and Simmonds [15] to obtain more appropriate values for the parameters in the potential for interactions between different molecular fragments, rather than by varying the parameters by trial and error. Indeed, as we saw in Sect. 5 the original parameters for the Gay-Berne potential, as proposed by Gay and Berne, were obtained for four Lennard-Jones sites and not for a realistic mesogen and so it seems sensible to use a more appropriate parameter set in future simulations to obtain more realistic properties.

The behaviour of the dimer model studied by Wilson [83] contrasts with that found by Luckhurst et al. [89] in their early attempt to simulate the behaviour of liquid crystal dimers by using Gay-Berne potentials. In this two GB(3.0, 5.0, 2, 1) mesogenic groups were linked together not via a chain of Lennard-Jones atoms but directly. That is, the point of attachment is within the ellipsoidal surface of each Gay-Berne mesogen and located on its symmetry axis. There is, however, a torsional and bond bending potential as in Eq. (49) with the parameters in the torsional potential set equal to those proposed by Ryckaert and Bellemans [90]. At a scaled number density of 0.27 the monomer GB(3.0, 5.0, 2, 1) undergoes a transition at a low temperature from the isotropic phase to a highly ordered phase which might be a smectic B or possibly a crystal. At the equivalent number density of 0.135 the dimer is found to form a nematic and a smectic A phase in which the layer spacing is about half the molecular length. In other words the smectic phase has an intercalated structure analogous to that found for real dimers [91]. The simulations also reveal the strong influence of the orientational order on the torsional distribution function and predicted by molecular field theories [92].

The GB-LJ hybrid models clearly offer a computationally achievable route for the study of flexible liquid crystals. The atomistic detail of the rigid mesogenic group can be replaced by the more generic Gay-Berne potential and the alkyl chain accounted for by a number of united atom Lennard-Jones sites. However, for these models to reproduce the behaviour observed in experimental systems, the potential must clearly be parametrised in a systematic way to obtain an appropriate model. It is also clear that, since the atomistic detail of the mesogen is lost, these models will not be useful for determining experimental data, such as the transition temperatures, accurately for which fully atomic detailed models (and very powerful computers) must be used. However, they should be a useful aid to study how the phase behaviour changes when generic features such as the length of the chain, its parity, or the angle of attachment are varied.

10
Biaxial Gay-Berne Molecules

Although the common nematic and smectic A phases have uniaxial symmetry, the molecules that form them are not cylindrically symmetric, although they are often assumed to have this high symmetry. Indeed, since no real mesogenic molecules are strictly uniaxial we may wonder what effect the molecular biaxiality has on the phase behaviour. It has been suggested that the high orientational order parameter and the large density jump at the nematic-isotropic transition given by, for example, the Onsager theory may be due to the approximation of a uniaxial molecular shape, since both these quantities are diminished for particles of biaxial shape [93, 94]. Perhaps more interestingly, biaxiality in the molecular shape and attractive interactions also introduces the possibility of a homogeneous fluid phase with biaxial symmetry, the biaxial nematic. In this phase, each molecular symmetry axis has a preferred direction of alignment, in contrast to the single unique direction of alignment in the uniaxial nematic. In order to study the behaviour of biaxial molecules in both uniaxial and biaxial phases, various extensions have been proposed to the Gay-Berne potential. However, we should note that there is still some doubt as to whether the biaxial nematic phase has been found for real thermotropic mesogens [95].

10.1
Single Site Biaxial Gay-Berne Models

One of the major benefits of the Gay-Berne potential is that it combines anisotropy in both repulsive and attractive forces in a single site potential. This leads to the efficient calculation of the potential energy or the forces and torques between a pair of molecules in a Monte Carlo or a molecular dynamics simulation, respectively. Thus, to investigate the effect of molecular biaxiality, there is a clear need to extend the Gay-Berne potential to biaxial particles. Berardi et al. [96] have done this by following the gaussian overlap procedure of Berne and Pechukas [11] to develop an appropriate range parameter for a pair of identical biaxial ellipsoidal molecules. The principle behind this generalisation is simple; the gaussian function representing each molecule is stretched by different amounts along three orthogonal molecular axes to obtain a biaxial object. The overlap between two such gaussian functions is then taken to give the interaction energy between a pair of molecules. The resulting potential can be expressed in terms of a range parameter which can be extracted and used in a potential of either Lennard-Jones form, as done by Berne and Pechukas [11] or shifted Lennard-Jones form as used by Gay and Berne [8]. Berardi et al. [96] also define a biaxial energy parameter analogous to that for the uniaxial case but which takes into account the different well depths for different arrangements of parallel molecules. Cleaver et al. [97] have also shown that a biaxial range parameter is available from the Perram-Wertheim contact function, normally used to test for overlap between hard biaxial ellipsoids [36]. Such extensions to the Gay-Berne potential clearly increases the number of parameters necessary to study a particular model; that is, from six (σ_s, σ_i, ε_s, ε_i, μ, ν) for the uniaxial model to eight (σ_x, σ_y, σ_z, ε_x, ε_y, ε_z, μ, ν) for the biaxial model. To reduce the number of parameters for the biaxial model, Ginzburg et al. [98] have proposed a method to calculate the attractive anisotropy directly from the shape anisotropy; they also show that this method is applicable to the uniaxial case. This method relies on the assumption that all of the attractive biaxiality is determined by the shape biaxiality and that the molecules are uniform, in the sense that all constituent parts interact with the same potential. Of course, these assumptions will not hold for molecules which are highly unsymmetric with heavy or polar groups at one end and not the other; however, this is also true of the uniaxial Gay-Berne potential, which provides one of the reasons for modifications such as the additions of electrostatic terms discussed in Sect. 7. The approach adopted by Ginzburg et al. [98] reduces the number of parameters by three for the biaxial model and two for the uniaxial case. This leads to a clear advantage over the formally correct model of Berardi et al. [96] for simulation studies since more complete investigations of the phase behaviour in which all parameters are varied become feasible. Of course, for simulations, this advantage is not so clear because very rarely can more than one parameter be varied because of the computational cost of determining the phase behaviour for each parameter set. In addition, it is likely that the additional parameters in the extended Gay-Berne potential of Berardi et al. [96] are necessary to ensure that the realistic features of the biaxial particle are retained. Despite the introduction of these models, to date no simulations using them have been published in which biaxial phases have been observed.

10.2
Multiple Site Biaxial Gay-Berne Models

An alternative approach to introducing molecular biaxiality is to build a multiple site molecule from a small number of Gay-Berne units similar to that employed to construct chiral molecules (see Sect. 8). This approach has been followed by Sarman [99] although primarily to test a director constraint algorithm for molecular dynamics simulations. The composite object is made up of nine oblate Gay-Berne interaction sites, placed in a straight line with their orientation axes parallel to each other but perpendicular to the long axis of the composite body. The Gay-Berne potential was altered to be of the form $1/R^{18}$ rather than $[1/R^{12}-1/R^6]$ to make the potential more short ranged. This does, of course, mean that the potential is now purely repulsive. However, the parameter κ' is retained although its physical interpretation is not so clear in this adjusted potential since there are no potential wells. The values for the parameters for the individual Gay-Berne units appear to have been selected somewhat arbitrarily in order to give a length-to-breadth-to-width ratio of 5:1:0.4; this shape is close, but by no means equal to, the optimal shape $(a, \sqrt{a}, 1)$ for hard biaxial ellipsoids needed to find a biaxial nematic phase [100]. This model is found to give a biaxial nematic phase directly from an isotropic phase on increasing the density and on lowering the density directly from the crystal. This phase was characterised by observing non-zero values for the usual uniaxial second rank orientational order parameter (see Eq. 9)

$$\overline{P_2} = \overline{Q_{00}^2} = \tfrac{1}{2}(\overline{3\cos^2\theta} - 1) \tag{50}$$

and a biaxial order parameter

$$Q_{22}^2 = \tfrac{1}{2}\overline{(1 + \cos^2 \theta) \cos 2\phi \cos 2\psi} - \overline{\cos \theta \sin 2\phi \sin 2\psi} \tag{51}$$

in which θ, ϕ and ψ are the Euler angles describing the orientation of a molecule relative to a laboratory based co-ordinate system containing the major and minor directors. Unfortunately, the structure of this phase was not investigated further.

11
Mixtures of Gay-Berne Molecules

11.1
Gay-Berne Mesogens in Mixtures

Many technological applications of liquid crystals, as in electro-optic display devices, are based on multicomponent mixtures. Such systems offer a route to the desired material properties which cannot be achieved simultaneously for single component systems. Mixtures also tend to exhibit a richer phase behaviour than pure systems with features such as re-entrant nematic phases [3] and nematic-nematic transitions possible. In this section, we describe simulations which have been used to study mixtures of thermotropic calamitic mesogens.

Until relatively recently, simulations of Gay-Berne mesogens have been confined to the study of single component systems. This is hardly surprising, since the dependence of the phase behaviour on the different parameters in the Gay-Berne potential is still not well-understood, although significant progress has been made in this area in recent years, as we have seen in Sect. 5. Another, more fundamental, reason for the lack of studies of mixtures is that the Gay-Berne potential in its original form only describes the interaction between two identical molecules, although we note that Berne and Pechukas [11] did give the range parameter for an atom-molecule potential. To model the interactions between a pair of dissimilar molecules, Lukac and Vesely used the Lorentz-Berthelot mixing rules [2] in their preliminary study of a binary Gay-Berne mixture [101]. Thus the orientational dependent distance and energy parameters in the original Gay-Berne potential (see Eqs. 2 and 4) were replaced by

$$\sigma_{AB}(\hat{\mathbf{u}}_i \hat{\mathbf{u}}_j \hat{\mathbf{r}}) = \tfrac{1}{2} \left[\sigma_{AA}(\hat{\mathbf{u}}_i \hat{\mathbf{u}}_j \hat{\mathbf{r}}) + \sigma_{BB}(\hat{\mathbf{u}}_i \hat{\mathbf{u}}_j \hat{\mathbf{r}}) \right], \tag{52}$$

and

$$\varepsilon_{AB}(\hat{\mathbf{u}}_i \hat{\mathbf{u}}_j \hat{\mathbf{r}}) = \left[\varepsilon_{AA}(\hat{\mathbf{u}}_i \hat{\mathbf{u}}_j \hat{\mathbf{r}}) \varepsilon_{BB}(\hat{\mathbf{u}}_i \hat{\mathbf{u}}_j \hat{\mathbf{r}}) \right]^{1/2}, \tag{53}$$

for a pair of molecules i and j of type A and B, respectively. Here $\sigma_{AA}(\hat{\mathbf{u}}_i \hat{\mathbf{u}}_j \hat{\mathbf{r}})$ is the range parameter for the pair of molecules calculated as if they are both of type A. Of course, if i and j are of the same type, then the usual Gay-Berne potential is used. However, for different types of molecules, that is, with either different shape or energy parameters, or both, then the mixing rules give an average interaction. As pointed out by Bemrose et al. [102, 103], this approach is not strictly appropriate for anisometric molecules, since the mixing rules fail to distinguish between the different T configurations for an AB pair; that is, the AB interaction is not the same as the BA interaction. However, this discrepancy is not expected to be too serious if the molecules are of similar dimensions, especially in the liquid crystalline phases, where parallel configurations of molecules are expected to be dominant and the T arrangement does not play a major role. Clearly, a more appropriate form of the potential is necessary for sufficiently different mesogens such as in mixtures of long and short rods or in the more extreme case of rods and discs, where the breadths as well as the lengths may differ significantly. Such a potential has been formulated by Cleaver et al. [104] to represent the interactions between dissimilar molecules and here we highlight the major features of the model.

11.1.1
The Generalised Gay-Berne Potential

In the original determination of the range parameter $\sigma(\hat{u}_i\hat{u}_j\hat{r})$ for identical molecules, Berne and Pechukas [11] considered each molecule to be represented by a gaussian cloud, stretched along the molecular symmetry axis to give it its anisotropic shape. The overlap between these clouds is taken to give the strength of the interaction, which can be written in terms of $\sigma(\hat{u}_i\hat{u}_j\hat{r})$. The range parameter can be extracted from the resulting potential and used in a more realistic Lennard-Jones-like function. Cleaver et al. [104] followed a similar method to determine the distance parameter $\sigma_{AB}(\hat{u}_i\hat{u}_j\hat{r})$ for dissimilar mesogens by considering the more general model in which the two gaussians are stretched by different amounts along the molecular symmetry axes. No restrictions, apart from the condition of uniaxiality, were placed on the scaling variables, so the potential can be employed to model any combination of oblate, prolate and spherical molecules. The range parameter for dissimilar molecules turns out to have a very similar form to that for the original Berne-Pechukas model,

$$\sigma_{AB}(\hat{\mathbf{u}}_i, \hat{\mathbf{u}}_j, \hat{\mathbf{r}}) = \sigma_0 \left[1 - \chi \left(\frac{\alpha^2(\hat{\mathbf{u}}_i.\hat{\mathbf{r}})^2 + \alpha^{-2}(\hat{\mathbf{u}}_j.\hat{\mathbf{r}})^2 - 2\chi(\hat{\mathbf{u}}_i.\hat{\mathbf{r}})(\hat{\mathbf{u}}_j.\hat{\mathbf{r}})(\hat{\mathbf{u}}_i.\hat{\mathbf{u}}_j)}{1 - \chi^2(\hat{\mathbf{u}}_i.\hat{\mathbf{u}}_j)^2} \right) \right]^{1/2},$$

(54)

where the additional parameter α reflects the difference between the dimensions of the pair of molecules and is given by

$$\alpha^2 = \left[\frac{(\sigma_{eA}^2 - \sigma_{sA}^2)(\sigma_{eB}^2 + \sigma_{sA}^2)}{(\sigma_{eB}^2 - \sigma_{sB}^2)(\sigma_{eA}^2 + \sigma_{sB}^2)} \right]^{1/2},$$

(55)

in which σ_{eA} and σ_{sA} are the length and breadth, respectively, of the molecules of species A. The parameter χ (see Eq. 3) must also be modified since the molecules no longer have the same dimensions; for unlike molecules, this is

$$\chi = \left[\frac{(\sigma_{eA}^2 - \sigma_{sA}^2)(\sigma_{eB}^2 + \sigma_{sB}^2)}{(\sigma_{eB}^2 - \sigma_{sA}^2)(\sigma_{eA}^2 + \sigma_{sB}^2)} \right]^{1/2} . \tag{56}$$

The range parameter is used in the shifted Lennard-Jones potential just as in the Gay-Berne potential. Since the well depth parameter in the Gay-Berne model was determined in a phenomenological way, the generalisation of this term is not so straightforward. However, Cleaver et al. [104] suggest a form which is modified by a parameter α', in the same manner as α enters the expression for the contact distance parameter. The addition of this parameter allows the energy for the two T configurations to differ. The parameter χ' is now a variable and is not related to the potential well depths in such a straightforward manner as in the original Gay-Berne model. Thus the relationship between χ' and α' to molecular features is not so simple as for χ and α. Indeed, these may now be thought to play a similar role as the two exponents μ and ν (see Eq. 6) in the sense that these are parameters which can be varied to give an appropriate fit to a more complicated model which the simulator requires to simplify. The use of a fitting procedure, such as that used by Luckhurst and Simmonds [15], is clearly essential to obtain appropriate parameters for the potential to be successful in modelling the behaviour of real mixtures, even at a generic level. This is especially important with the addition of two more parameters since the determination of phase diagrams in which the adjustable parameters occurring in the potential are modified individually is clearly impossible. Indeed, even for the original model for single component systems, this is proving to be a formidable task.

11.1.2
A Binary Mixture

Bemrose et al. [102, 103] have performed constant NVE and NPT simulations of a binary mixture of Gay-Berne mesogens, where the components differ in both their energy and shape anisotropy. We shall concentrate on the results of their NPT simulations, which were performed by varying the concentrations of the two species. The use of the constant pressure ensemble was necessary since it is difficult to define a meaningful density for a Gay-Berne mixture with which to compare two systems of different composition because the effective volume of a Gay-Berne molecule is not well-defined and so it is not straightforward to ensure that the systems are simulated at the same relative densities or packing fractions. For example, if a constant box size is used, the packing fractions of two systems with different concentrations will be different, although the number densities will be the same, and so it would be difficult to compare the results of two simulations. The two species studied by Bemrose et al. had length-to-breadth ratios (A) 3.5:1.0 and (B) 3.0:1.0 and were both of the same width; details of the other parameters are given in the original paper [102]. Two simulations were run to determine the phase behaviour for pure A and pure B. These were followed by simulations of mixtures at intermediate mole fractions. Large density and order parameter jumps were found for each series as the system was cooled; this corresponds to the transition at constant pressure from the isotropic to a smectic B phase. This phase behaviour is similar to that for pure GB(3.0, 5.0, 2, 1) at constant pressure [43]. The transition temperature, T_{SmBI}, is found to increase linearly with the mole fraction of A in accord with its greater anisotropy. Similarly, the smectic layer spacing is found to increase although this appears not to be as linear as the behaviour observed for T_{SmBI}. Since the addition of small quantities of the second component gradually increases the layer spacing, this implies that there is no demixing in the system although pair distribution functions should be calculated to confirm this point (see Sect. 3). However, it is found that the layer spacing for the mixture with 25% A does not fit the smooth behaviour and it is proposed that this is due to an extended nematic region. The justification for this conclusion is unclear since it seems to contradict the behaviour of the pure system of type B. Indeed, Bemrose et al. [102] also present data which suggests that the pure B system has a nematic phase but only the SmB-I transition temperature is shown in their main phase diagram.

This study of Gay-Berne mixtures indicates that the generalised Gay-Berne potential should provide a promising route with which to investigate liquid crystal mixtures. However, more work is necessary both to locate and understand the structure of the liquid crystalline regions exhibited by these models.

11.2
Other Mixtures

The study of mixtures is not confined to two mesogenic components. Thus Emsley and his colleagues have investigated dilute solutions of benzene [86] and hexane [105] dissolved in a Gay-Berne mesogenic solvent to clarify the orientational order of both solutes as well as the conformational distribution for hexane. In each case, the solute molecule is simulated with a united atom model and the solvent is modelled by the GB(3.0, 5.0, 2, 1) potential. Rather than use the molecule-atom range parameter proposed by Berne and Pechukas [11] to calculate the interaction between the Gay-Berne solvent and the united atoms of the solute, the authors calculate this interaction by replacing the Gay-Berne unit by a linear array of four Lennard-Jones sites. Whilst this gives a reasonable fit for certain configurations, it does mean that the united atoms of the solute molecule do not feel the smooth energy contours of a true Gay-Berne particle, but rather the rippled ones of the four-site model. In addition, the shape of the four-site model is essentially that of a spherocylinder and so quite different to the ellipsoidal shape of a Gay-Berne molecule. However, these details aside, the model is found to give good agreement with experimental NMR data for simulated proton-proton dipolar couplings in the nematic phase. For example, the conformation of a hexane molecule in the isotropic phase at low density is found to be very similar to that in a nematic phase at twice the density. Whilst the nematic phase should favour conformations such as the all-trans conformation, the large increase in density should favour more gauche links; thus these two factors seem to balance each other. However, on cooling at constant density to the smectic B phase, there is a considerable enhancement of the all-trans conformation, indicating that there is considerable solute orientational ordering, which leads to the elongated structure, induced by the solvent and in agreement with molecular field theory [106]. To understand fully the effect of solute-solvent interactions it is clear that there is still further work to do and many more state points should be studied to try to decouple the effects of density and temperature. However, this is clearly a novel use for the Gay-Berne potential which opens up a whole field of possibilities to explore the structure of solutes in liquid crystalline systems. Also, as noted by Alejandre et al. [105], even though the system appears to be dramatically simplified, and possibly inconsistent, the fact that experimental data such as the NMR dipolar coupling constants can be calculated from the simulations which match those determined in real experiments is particularly reassuring and shows that these simple models have much to offer.

We also note that mixtures can be modelled in other ways. For example, Bates and Luckhurst [27] have studied mixtures of quadrupolar discs in which the components differ in the sign of their quadrupole moments. These simulations are discussed in Sect. 7.2.

12
Gay-Berne Mesogens at Interfaces and in Confined Geometries

In the previous sections, we have seen how computer simulations have contributed to our understanding of the microscopic structure of liquid crystals. By applying periodic boundary conditions preferably at constant pressure, a bulk fluid can be simulated free from any surface interactions. However, the surface properties of liquid crystals are significant in technological applications such as electro-optic displays. Liquid crystals also show a number of interesting features at surfaces which are not seen in the bulk phase and are of fundamental interest. In this final section, we describe recent simulations designed to study the interfacial properties of liquid crystals at various types of interface. First, however, it is appropriate to introduce some necessary terminology.

In a bulk liquid crystal, the free energy does not depend on the orientation of the director and so, in the absence of any constraining forces, the liquid crystal can align in any direction. However, at an interface the translational symmetry is broken and a specific direction is favoured. This direction will depend on the nature of the interactions between the molecules and the surface. This phenomenon is known as surface anchoring and the direction of director alignment which minimises the free energy is known as the easy axis. Depending on the angle, θ, between the director and the surface normal, we can define homeotropic ($\theta = 0$), tilted ($0 \le \theta \le \pi/2$) and planar ($\theta = \pi/2$) anchoring [107, 108]. If the bulk liquid phase is isotropic, rather than nematic, the molecules at the interface may still be aligned in a specific direction and this alignment may or may not propagate into the bulk. Thus, if we consider the isotropic-vapour interface of a nematogen, for example, it is possible to observe a thin layer of nematic between the two bulk phases; the nematic is said to wet the surface of the isotropic.

12.1
The Nematic-Solid Interface

Stelzer et al. [109] have studied the case of a nematic phase in the vicinity of a smooth solid wall. A distance-dependent potential was applied to favour alignment along the surface normal near the interface; that is, a homeotropic anchoring force was applied. The liquid crystal was modelled with the GB(3.0, 5.0, 2, 1) potential and the simulations were run at temperatures and densities corresponding to the nematic phase. Away from the walls the molecules behave just as in the bulk. However, as the wall is approached, oscillations appear in the density profile indicating that a layered structure is induced by the interface, as we can see from the snapshot in Fig. 19. These layers are

Fig. 19. A snapshot of a configuration showing a surface-induced smectic A phase near a smooth wall for the GB(3.0, 5.0, 2, 1) mesogen

found to have in-plane disorder and so the induced phase was characterised as a smectic A. We recall that in the bulk the GB(3.0, 5.0, 2, 1) mesogen does not exhibit a smectic A phase. Thus this simulation indicates that a solid surface can induce phase behaviour not seen in the bulk. The induced behaviour is found to be temperature dependent; thus, as the temperature is lowered, so the number of induced layers increases.

Analogous surface-induced ordering effects have been observed for the same Gay-Berne mesogen by Gruhn and Schoen [110] but with the two confining surfaces constructed from atoms. These are arranged on the (100) face of a face-centred-cubic lattice with atoms for both surfaces in register, each atom interacting with a Gay-Berne particle via a scaled and shifted 12–6 potential (see Eqs. 1 and 5). The contact distance is written as [111]

$$\tilde{\sigma}(\hat{\mathbf{u}}_i\hat{\mathbf{r}}) = \tilde{\sigma}_s\{1 - \eta(\hat{\mathbf{u}}_i.\hat{\mathbf{r}})\}^{-1/2}, \tag{57}$$

where the shape anisotropy parameter η is given by

$$\eta = 1 - (\tilde{\sigma}_e/\tilde{\sigma}_s)^{-2}. \tag{58}$$

Here, the tilde indicates the parameter for the combination of a rod-like molecule and an atom so $\tilde{\sigma}_e$ denotes the contact distance when the atom is at the end of the molecule and $\tilde{\sigma}_s$ that when it is at the side. The analogous expression used for the well depth is

$$\tilde{\varepsilon}(\hat{\mathbf{u}}_i\hat{\mathbf{r}}) = \tilde{\varepsilon}_s\{1 - \eta'(\hat{\mathbf{u}}_i.\hat{\mathbf{r}})^2\}^2, \tag{59}$$

where

$$\eta' = 1 - (\tilde{\varepsilon}_s/\tilde{\varepsilon}_e). \tag{60}$$

By careful selection of the parameter η' it is possible to favour either homeotropic or planar alignment of a Gay-Berne molecule at the surface. The system was studied using grand canonical ensemble Monte Carlo simulations because it is intended to model an open system such as those investigated in a surface-force apparatus. The chemical potential and temperature were set at values corresponding to the isotropic phase for bulk GB(3.0, 5.0, 2, 1) so that the influence of the surface constraints on the long range structure could be easily seen. For homeotropic alignment induced by setting η' to a value greater than unity a monolayer of Gay-Berne molecules is created at the surface with a relatively high degree of orientational and transitional order. In addition the molecules tend to be parallel to the layer normal, and presumably the atomic surface acts as a template for this induced layer. The remaining Gay-Berne molecules in this thin film are found to remain in the disordered phase for of the system.

Similar behaviour has been observed by Emerson and Zannoni [112] in their simulations of polymer dispersed liquid crystal droplets where the solid surfaces are necessarily curved. The walls of the spherical cavity were modelled by constructing a rigid shell of Gay-Berne molecules in fixed positions and orientations analogous to the atomic surface used by Gruhn and Schoen [110]. In order to mimic radial boundary conditions, the molecules composing the surface were oriented with their symmetry axes directed towards the centre of the droplet. To ensure a strong surface anchoring potential, the wall-fluid interactions were modelled with the GB(3.0, 5.0, 1, 3) potential, which strongly favours parallel arrangements. Thus the end-to-end arrangement is more favoured than the T-configuration. The internal fluid-fluid interactions were modelled with both the GB(3.0, 5.0, 2, 1) and the GB(3.0, 5.0, 1, 3) potentials. In small droplets of scaled radius $r^* < 10$ containing the GB(3.0, 5.0, 2, 1) mesogen, a single radial shell is observed to form at the edge of the droplet. This shell is similar to the induced smectic layer seen in the simulations of Stelzer et al. [109], as well as by Gruhn and Schoen [110] with the obvious difference that here it is curved. For a larger droplet of scaled radius 20, further radial shells are observed to form inside the first and these start to propagate towards the centre of the droplet (as we can see from Fig. 20). We may wonder why the different size droplets containing the same fluid behave differently. It seems that the large droplet behaves in a similar way to the flat interface and so the solid wall can induce the formation of a number of layers. This is presumably because the deformation of the second and third layers due to the shape of the cavity are not prohibitive to their formation. In contrast, for the smaller droplets, the penalty which must be overcome to form further layers with a large curvature is too great and so a normal smectic-like domain forms in the middle of the outer curved layer. Different behaviour is found if the internal fluid interactions are modelled with the GB(3.0, 5.0, 1, 3) potential. As a result of the enhanced interactions between the fluid molecules, further shells propagate towards the centre even in a small droplet. We note that these droplets are extremely small; Emerson and Zannoni [112] calculate the diameter of the larger droplet to be less than 0.01 μm, whilst typical droplets in real polymer dispersed liquid crystal systems have diameters of 2–3 μm. However, they clearly show an interesting feature which is at variance with the hedgehog model for radially aligned droplets, in which a point is expected at the centre. Thus, the curvature appears to be too great to form radially aligned shells. Of course, this is for radial smectic-like shells, but we expect that a similar behaviour at high curvature will also hold for nematogens. The model introduced by Emerson and Zannoni [112] should also be easily extended to droplets with other boundary conditions by altering the alignment of the molecules in the fixed outer wall.

12.2
The Nematic-Vapour Interface

The alignment of the director at the nematic free surface of real systems is not found to exhibit universal behaviour. Depending on the mesogen, homeotropic, tilted and planar anchoring have been observed. Clearly, to study this interface in a simulation a potential which exhibits a nematic phase in co-

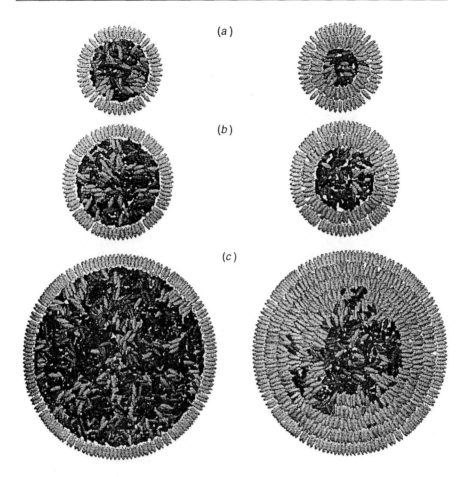

Fig. 20a–c. Equatorial sections through polymer dispersed Gay-Berne liquid crystal droplets of various radii at high ($T^* = 3.0$) (*left*) and low ($T^* = 0.6$) (*right*) scaled temperatures for droplets with scaled radii of: **a** 7.1; **b** 10.0; **c** 20.0

existence with the vapour is necessary. Unfortunately, the original parametrisation of the Gay-Berne potential does not satisfy this requirement, as we saw
in Sect. 5. However, the original version of the potential does exhibit isotropic-
vapour co-existence and the I-V interface has been studied [113]. In these
simulations, even though the bulk of the fluid film is isotropic, at the surface
the molecules are observed to align perpendicular to the interface.

By systematically modifying the well depth ratio κ' whilst keeping the
original values for the other parameters ($\kappa = 3.0$, $\mu = 2$ and $\nu = 1$), de Miguel
et al. [45] have shown that nematic-vapour co-existence is possible if κ' is
reduced to below 2.0. However, this low value seems somewhat unrealistic;
indeed we recall that the value obtained by Luckhurst and Simmonds [15] for
this parameter in order for the potential to be appropriate for real mesogens
was close to 40. This low value for κ' also essentially rules out the possibility of
observing a bulk smectic phase, since the strength of interaction for pairs of
molecules in the side-to-side arrangement is significantly reduced compared to
other parallel arrangements. However, this apparently, unrealistic feature of
the potential aside, the nematic-vapour interface has now been studied for two
different parametrisations ($\kappa' = 1.0$ [114] and 1.25 [115]). The reduction of κ'
from 5.0 to these lower values is found to stabilise planar anchoring at the
isotropic-vapour interface in contrast to the perpendicular alignment observed
for the higher value. Thus in both simulations, the molecules are found to align
parallel to the interface, even though the bulk phase is isotropic. Nematic
wetting is also reported for GB(3.0, 1.0, 2, 1) and thus the molecules in the
interfacial region tend to align in a specific direction. This was not observed in
the simulations of GB(3.0, 1.25, 2, 1) in which the orientations of the molecules
are isotropically distributed throughout the plane. The reason why the two
models should differ in this respect is not clear, given their similarity. In
addition, the nematic-vapour interface is also possible for these two models
and planar anchoring is observed in each case. Thus, the anchoring direction
at the vapour interface does not change on going from the isotropic to the
nematic phase. As we noted in Sect. 5, since the value of κ' is reduced to just
above unity, this rules out the possibility of studying the smectic-vapour
interface for these models, because there is no preference for the molecules to
align side-by-side, since the energy parameter is essentially constant for
parallel molecules.

The nematic-vapour interface has also been studied by simulating free
droplets [116] composed of GB(3.0, 1.25, 2, 1) molecules. Above the nematic-
isotropic transition, the droplets are found to be, on average, spherical. This
spherical shape is preserved on cooling just below the nematic-isotropic
transition, although the inside of the droplet becomes weakly aligned.
However, on further cooling, the droplets are found to become elongated as
we can see in Fig. 21. This elongation presumably occurs because the
molecules within the droplet tend to align in a specific but arbitrary direction.
In contrast, the molecules at the interface tend to align parallel to the surface.
This different alignment by the surface causes frustration and so, to relieve
this, the droplet becomes deformed. The GB(3.0, 5.0, 2, 1) mesogen also shows
interesting free droplet behaviour. We recall that this model exhibits

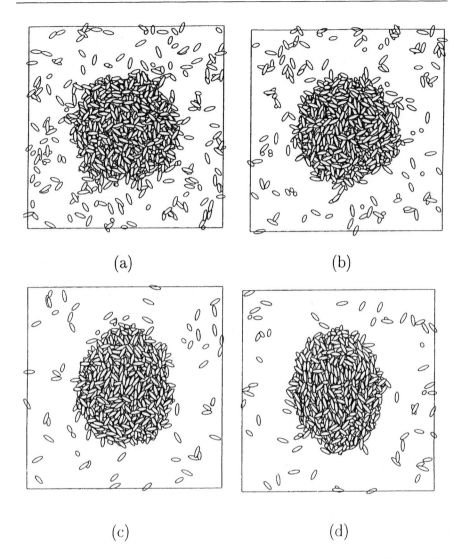

(a)   (b)

(c)   (d)

Fig. 21a–d. Free nematic droplets of GB(3.0, 1.25, 2, 1) at scaled temperatures, T^*, of: **a** 0.56; **b** 0.54; **c** 0.52; **d** 0.50. T^*_{NI} for the bulk fluid is approximately 0.55 [45]

homeotropic anchoring and so we would expect that the molecules in a free droplet of the mesogen GB(3.0, 5.0, 2, 1) would tend to align perpendicular to the surface. Thus, it would seem that a droplet composed of radial shells would be most favourable but, as Emerson and Zannoni found [112], the energy penalty to form highly curved shells in small droplets may be too great.

This is also observed to be the case for free droplets [116]. Indeed, simulations started from isotropic droplets below the smectic B-isotropic transition form cylindrical rather than spherical droplets; these are apparent in Fig. 22. In this way, the molecules can align in parallel layers with the

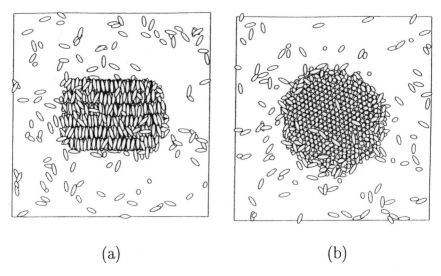

(a)   (b)

Fig. 22a,b. Snapshots of a smectic droplet formed by GB(3.0, 5.0, 2, 1) taken from: **a** the side; **b** the top

molecules in the top and bottom layers exhibiting homeotropic anchoring. The molecules at the curved surface of the cylinder are found to be attached relatively weakly to the droplet and are often seen to leave and later rejoin the droplet in molecular dynamics simulations. In contrast, evaporation from the top or bottom of the droplet is rare.

12.3
The Nematic-Isotropic and Smectic A-Isotropic Interfaces

As a nematic liquid crystal is heated to form an isotropic liquid, it undergoes a weak first order transition [3]. If a temperature gradient is applied to a suitably prepared system, a flat interface can exist between the two phases [117]. This method provides a technique with which to study a liquid-liquid interface, in which the material is the same on both sides of the interface, but the phase is different. The lack of simulation studies of the nematic-isotropic interface is probably due to the difficulty of both generating and maintaining an interface in the relatively small size systems presently accessible. A simple method has been introduced to overcome this problem by Bates and Zannoni [118], based on the preparation of a suitable temperature inhomogeneity in the system. In this method, one half of the simulation box is maintained at a temperature slightly above the transistion, whilst the other half is slightly below. This method has been used to study both the nematic-isotropic [118, 119] and smectic A-isotropic [119] interfaces. Since the interface between a liquid crystal and its isotropic phase is typically rather broad, a large system size is necessary to ensure that the two interfaces studied in a simulation box do not

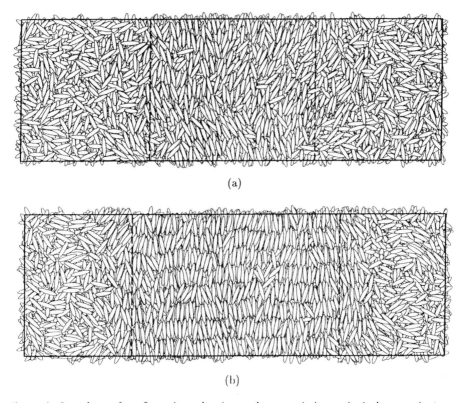

(a)

(b)

Fig. 23a,b. Snapshots of configurations showing: **a** the nematic-isotropic; **b** the smectic A-isotropic interfaces. The Gibbs dividing surfaces are indicated by the *dashed lines*

interfere with each other; two interfaces must be studied because of the periodic boundary conditions. These simulations were, therefore, performed on 12,960 [118] and 14,256 [119] molecules. In both cases, the molecules tend to align, on average, perpendicular to the layer normal, as shown in Fig. 23. Thus the orientational anchoring at the interface is planar rather than homeotropic or tilted. In both systems, there is an offset between the Gibbs dividing surfaces or the orientational order parameter profile and the density profile; the Gibbs diving surface can be viewed as the centre of the interface, determined for a particular parameter profile. The explanation for this is that, even in the orientationally disordered isotropic phase, it takes a few molecular widths for the orientational correlations present between the molecules to decay. Thus the orientational order penetrates the low density side of the interface. For the smectic A-isotropic system, a similar offset is found between the orientational order parameter profile $\bar{P}_2(z^\star)$ and the translational order parameter $\tau(z^\star)$ profile. Thus, the orientational order is propagated towards the isotropic phase but the translational order is not. This results in a narrow region at the interface between the smectic A and isotropic phases which is

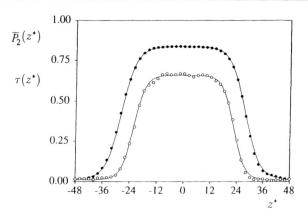

Fig. 24. The orientational, $\bar{P}_2(z^*)$, and translational, $\tau(z^*)$, order parameter profiles for the smectic A-isotropic interface. The offset between the two profiles of about five molecular widths supports the observation of a narrow region with a nematic structure between the smectic A and isotropic phases

nematic-like; this is shown in Fig. 24. Thus, for this model, nematic wetting is observed at the smectic A-isotropic interface.

13
References

1. Hansen JP, McDonald IR (1986) Theory of simple liquids, 2nd edn. Academic Press, New York
2. Allen MP, Tildesley DJ (1987) Computer simulation of liquids. Clarendon Press, Oxford
3. Collings PJ, Hird M (1997) Introduction to liquid crystals. Taylor and Francis, London
4. Lebwohl PA, Lasher G (1972) Phys Rev A 6: 426
5. Vieillard-Baron J (1974) Mol Phys 28: 809
6. Komolkin AV, Molchanov Yu V, Yakutseni PP (1989) Liq Cryst 6: 39
7. Corner J (1948) Proc R Soc Lond A 192: 275
8. Gay JG, Berne BJ (1981) J Chem Phys 74: 3316
9. Adams DJ, Luckhurst GR, Phippen RW (1987) Mol Phys 61: 1575
10. Luckhurst GR, Stephens RA, Phippen RW (1990) Liq Cryst 8: 451
11. Berne BJ, Pechukas P (1972) J Chem Phys 56: 4123
12. Stone AJ (1979) In: Luckhurst GR, Gray GW (eds) Molecular physics of liquid crystals, chap 2. Academic Press, New York
13. (a) Walmsley SH (1977) Chem Phys Lett 49: 320; (b) Tsykalo AL, Bagmet AD (1978) Mol Cryst Liq Cryst 46: 111
14. Bates MA, Luckhurst GR (1998) J Chem Phys (submitted)
15. Luckhurst GR, Simmonds PSJ (1993) Mol Phys 80: 233
16. Emerson APJ, Luckhurst GR, Whatling SG (1994) Mol Phys 82: 113
17. Bates MA, Luckhurst GR (1996) J Chem Phys 104: 6696
18. Zannoni C (1994) In: Luckhurst GR, Veracini CA (eds) Molecular dynamics of liquid crystals, chap 2. Kluwer, Dordrecht
19. Wojtowicz PJ (1974) In: Priestly EB, Wojtowicz PJ, Sheng P (eds) Introduction to liquid crystals, chap 7. Plenum, New York
20. Kventsel GF, Luckhurst GR, Zewdie HB (1985) Mol Phys 56: 589

21. Veerman JAC, Frenkel D (1992) Phys Rev A 45: 5632
22. Emerson APJ, Hashim R, Luckhurst GR (1992) Mol Phys 76: 241
23. Berardi R, Emerson APJ, Zannoni C (1993) J Chem Soc Faraday Trans 89: 4069
24. Leadbetter AJ (1987) In: Gray GW (ed) Thermotropic liquid crystals, chap 1. Wiley, New York
25. Strandberg KJ (1992) In: Strandberg KJ (ed) Bond orientational order in condensed matter systems, chap 2. Springer, Berlin Heidelberg New York
26. Hashim R, Luckhurst GR, Romano S (1984) Mol Phys 53: 1535
27. Bates MA, Luckhurst GR (1998) Liq Cryst 24: 229
28. Zannoni C (1979) In: Luckhurst GR, Gray GW (eds) Molecular physics of liquid crystals, chap 3. Academic Press, New York
29. Gray CG, Gubbins KE (1984) Theory of molecular fluids, vol 1 – Fundamentals. Clarendon Press, Oxford
30. (a) Leadbetter AJ (1979) In: Luckhurst GR, Gray GW (eds) Molecular physics of liquid crystals, chap 13. Academic Press, London; (b) Doucet J (1979) In: Luckhurst GR, Gray GW (eds) Molecular physics of liquid crystals, chap 14. Academic Press, London
31. Guinier A, Fournet G (1955) Small angle scattering of X-rays. Wiley, New York
32. Hamley IW (1990) PhD Thesis, University of Southampton
33. (a) Bates MA (1996) PhD Thesis, University of Southampton; (b) Bates MA, Luckhurst GR (in preparation)
34. Andersen HC, Chandler D, Weeks JD (1976) Adv Chem Phys 34: 105
35. Longuet-Higgins HC, Widom B (1964) Mol Phys 8: 549
36. Allen MP, Evans GT, Frenkel D, Mulder BM (1993) Adv Chem Phys 86: 1
37. Bolhuis P, Frenkel D (1997) J Chem Phys 106: 666
38. (a) Frenkel D, Ladd AJC (1984) J Chem Phys 81: 3188; (b) Frenkel D, Mulder BM, McTague JP (1984) Phys Rev Lett. 52: 287; (c) Frenkel D, Mulder BM (1985) Mol Phys 55: 1171
39. (a) Zarragoicoechea GJ, Levesque D, Weis JJ (1991) Mol Phys 74: 629; (b) Zarragoicoechea GJ, Levesque D, Weis JJ (1992) Mol Phys 75: 989
40. Frenkel D (1989) Liq Cryst 5: 929
41. Frenkel D (1987) Mol Phys 60: 1
42. Bates MA, Luckhurst GR (unpublished results)
43. Hashim R, Luckhurst GR, Romano S (1995) J Chem Soc Faraday Trans. 91: 2141
44. de Miguel E, Rull LF, Chalam MK, Gubbins KE (1991) Mol Phys 74: 405
45. de Miguel E, Martin del Rio E, Brown JT, Allen MP (1996) J Chem Phys 105: 4234
46. Luckhurst GR, Simmonds PSJ (1993) Mol Phys 80: 233
47. Bates MA, Luckhurst GR (1997) Chem Phys Lett 281: 193
48. McColl JR (1972) Phys Lett. A 38: 55
49. Luckhurst GR (1979) In: Luckhurst GR, Gray GW (eds) Molecular physics of liquid crystals, chap 4. Academic Press, New York
50. de Gennes PG (1974) Physics of liquid crystals. Clarendon Press, Oxford, p 42
51. Luckhurst GR, Setaka M, Zannoni C (1974) Mol Phys 28: 49
52. (a) McMillan WL (1971) Phys Rev A 4: 1238; (b) McMillan WL (1972) Phys Rev A 6: 936
53. Kventsel GF, Luckhurst GR, Zewdie HB (1985) Mol Phys 56: 589
54. van Roij R, Bolhuis P, Mulder B, Frenkel D (1995) Phys Rev E 52: 1277
55. Bates MA, Luckhurst GR (in preparation)
56. Vörlander D (1923) Zeit Phys Chem 105: 211
57. (a) Chandrasekhar S, Sadashiva BK, Suresh KA (1977) Pramana 9: 471; (b) Chandrasekhar S (1983) Philos Trans R Soc Lond A 309: 93
58. Chandrasekhar S (1993) Liq Cryst 14: 3
59. Eppenga R, Frenkel D (1984) Mol Phys 52: 1303
60. Bates MA, Frenkel D (1998) Phys Rev E 57: 4824
61. Satoh K, Mita S, Kondo S (1996) Liq Cryst 20: 757
62. Satoh K, Mita S, Kondo S (1996) Chem Phys Lett 255: 99
63. Gwózdz E, Bródka A, Pasterny K (1997) Chem Phys Lett 267: 557

64. Berardi R, Orlando S, Zannoni C (1996) Chem Phys Lett 261: 357
65. Frenkel D (1993) In: Allen MP, Tildesley DJ (eds) Computer simulation in chemical physics. Kluwer, Dordrecht
66. Berardi R, Orlando S, Zannoni C (1997) J Chem Soc Faraday Trans 93: 1493
67. McMillan WL (1973) Phys Rev A 8: 1921
68. Wulf A (1975) Phys Rev A 11: 365
69. Bacchiochi C, Bates MA, Luckhurst GR (in preparation)
70. Praefcke K, Singer D, Kohne B, Ebert M, Liebmann A, Wendorff JH (1991) Liq Cryst 10: 147
71. Patrick CR, Prosser GS (1960) Nature 187: 1021
72. (a) Van der Meer BW, Vertogen G, Dekker AJ, Ypma JGJ (1976) J Chem Phys 65: 3935; (b) Lin-Liu YR, Yu Ming Shi, Chia-Wei Woo, Tan HT (1976) Phys Rev A 14: 445; (c) Lin-Liu YR, Yu Ming Shi, Chia-Wei Woo, Tan HT (1977) Phys Rev A 15: 2550
73. Van der Meer BW, Vertogen G (1979) Luckhurst GR, Gray GW (eds) Molecular physics of liquid crystals, chap 6. Academic Press, New York
74. Memmer R, Kuball H-G, Schönhofer A (1993) Liq Cryst 15: 345
75. (a) Tsykalo AL (1979) Zh Fiz Khim 53, 2528; translation (1979) Russ J Phys Chem 53: 1143)
76. (a) Memmer R, Kuball H-G, Schönhofer A (1993) Ber Bunsen Ges Phys Chem 97: 1193; (b) Allen MP, Masters AJ (1993) Mol Phys 79: 227; (c) Yoneya M, Berendsen HJC (1994) J Phys Soc Jpn 63: 1025; (d) Saslow WM, Gabay M, Zhang WM (1992) Phys Rev Lett 68: 3627
77. Luckhurst GR, Romano S, Zewdie HB (1996) J Chem Soc Faraday Trans 92: 1781
78. Stegemeyer H, Blümel Th, Hiltrop K, Onusseit H, Porsch F (1986) Liq Cryst 1: 3
79. Memmer R, Kuball H-G, Schönhofer A (1995) Liq Cryst 19: 749
80. Gottarelli G, Spada GP, Varech D, Jaques J (1986) Liq Cryst 1: 29
81. McBride C, Wilson MR, Howard JAK (1998) Mol Phys 93: 955
82. La Penna G, Catalano D, Veracini CA (1996) J Chem Phys 105: 7097
83. Wilson MR (1997) J Chem Phys 107: 8654
84. Luckhurst GR (1995) MacroMol Symp 96: 1
85. Allen MP (1995) In: Baus M, Rull LF, Ryckaert J-P (eds) Observation, prediction and simulation of phase transitions in complex fluids. Kluwer, Dordrecht
86. Palke WE, Emsley JW, Tildesley DJ (1994) Mol Phys 82: 177
87. Emsley JW, Luckhurst GR, Shilstone GN, Sage I (1984) Mol Cryst Liq Cryst Lett 102: 223
88. Date RW, Imrie CT, Luckhurst GR, Seddon JM (1992) Liq Cryst 12: 203
89. Luckhurst GR, Simmonds PSJ, Tildesley DJ (unpublished results)
90. Ryckaert J-P, Bellemans A (1978) J Chem Soc Faraday Disc 66: 95
91. Hogan JL, Imrie CT, Luckhurst GR (1988) Liq Cryst 3: 645
92. Ferrarini A, Luckhurst GR, Nordio P-L, Roskilly SJ (1993) Chem Phys Lett 214: 409
93. Gelbart WM, Barboy B (1980) Acc Chem Res 13: 290
94. Tjipto-Margo B, Evans GT (1991) J Chem Phys 94: 4546
95. Hughes JR, Kothe G, Luckhurst GR, Malthête J, Neubert ME, Shenouda I, Timimi BA, Tittelbach M (1998) J Chem Phys 107: 9252
96. Berardi R, Fava C, Zannoni C (1995) Chem Phys Lett 236: 462
97. Cleaver DJ, Care CM, Allen MP, Neal MP (1996) Phys Rev E 54: 559
98. Ginzburg VV, Glaser MA, Clark NA (1997) Chem Phys Lett 214: 253
99. Sarman S (1996) J Chem Phys 104: 342
100. (a) Allen MP (1990) Liq Cryst 8: 499; (b) Camp PJ, Allen MP (1997) J Chem Phys 106: 6681
101. Lukac R, Vesely FJ (1995) Mol Cryst Liq Cryst 262: 533
102. Bemrose RA, Care CM, Cleaver DJ, Neal MP (1997) Mol Phys 90: 625
103. Bemrose RA, Care CM, Cleaver DJ, Neal MP (1997) Mol Cryst Liq Cryst 299: 27
104. Cleaver DJ, Care CM, Allen MP, Neal MP (1996) Phys Rev E 54: 559
105. Alejandre J, Emsley JW, Tildesley DJ, Carlson P (1994) J Chem Phys 101: 7027
106. Emsley JW, Luckhurst GR (1980) Mol Phys 41: 19

107. Sluckin TJ, Poniewierski A (1986) In: Croxton CA (ed) Fluid interfacial phenomena. Wiley, New York, pp 215–253

108. Jérome B (1991) Rep Prog Phys 54: 391

109. Stelzer J, Longa L, Trebin HR (1997) Phys Rev E 55: 7085

110. Gruhn T, Schoen M (1998) Mol Phys 93: 681

111. Gruhn T, Schoen M (1997) Phys Rev E 55: 2861

112. Emerson APJ, Zannoni C (1995) J Chem Soc Faraday Trans 91: 3441

113. Martin del Rio E, de Miguel E, Rull LF (1995) Physics A 213: 138

114. Martin del Rio E, de Miguel E (1997) Phys Rev E 55: 2916

115. Emerson APJ, Faetti S, Zannoni C (1997) Chem Phys Lett 271: 241

116. Bates MA, Emerson APJ, Zannoni C (unpublished results)

117. Bechhoefer J, Simon AJ, Libchaber A, Oswald P (1989) Phys Rev A 40: 2042

118. Bates MA, Zannoni C (1997) Chem Phys Lett 280: 40

119. Bates MA (1998) Chem Phys Lett 288: 209

Crystal Structures of LC Mesogens

W. Haase, M. A. Athanassopoulou

Darmstadt University of Technology, Institute of Physical Chemistry, Petersenstr. 20, 64287 Darmstadt, Germany
Email: *d54d@hrzpub.tu-darmstadt.de*

During the last two decades, the number of publications dealing with X-ray structure analyses of mesogens in the solid state has increased enormously. A comprehensive review of the crystal structures of mesogenic compounds is presented here.

Keywords: Liquid crystals, Crystal structures, Crystal packing, Conformation

1
Introduction

Liquid crystals form a state of matter intermediate between the ordered solid and the disordered liquid. These intermediate phases are called mesophases. In the crystalline state the constituent molecules or ions are ordered in position and orientation, whereas in the liquid state the molecules possess no positional and orientational ordering. Liquid crystals combine to some extent the properties of both the crystalline state (optical and electrical anisotropy) and the liquid state (fluidity).

X-ray diffraction is a very valuable tool for studying the structural features of crystals and liquid crystals because the X-ray wavelength is of the same order of magnitude as the interatomic and intermolecular distances. From X-ray data of single crystals, it is possible to obtain information about the crystal packing, the conformation of the molecules, and the molecular arrangement. A detailed knowledge of the structural features of mesogenic materials in their crystalline phases may lead to a better understanding of their structural arrangements in the mesophase. Therefore, the structure determinations of mesogenic compounds in the crystalline state have been established and reported by many authors.

In 1933, Bernal and Crowfoot [1] presented the first work on crystal structures of liquid crystals. In this pioneering work, the crystal structures of p-azoxyanisole and ethyl p-ethoxy-benzylideneamino cinnamate were described. These structural analyses provided support for the idea of a precursor for the nematic phase forming a parallel imbricated arrangement of molecules without translational restrictions and a precursor for the smectic phase forming a layered arrangement in the solid. Surprisingly, there were no reports on crystal structures of mesogenic compounds between 1933 and 1967. Galigné and Falgueirettes [2] presented the crystal structure of anisaldehyde azine in 1968. In the early seventies Krigbaum et al. re-examined the crystal structure of p-azoxyanisole [3] and reported also on the crystal structure of the smectogenic ethyl p-azoxybenzoate [4]. At the same time Galigné [5] described the crystal structure of the nematogenic 4,4′-azodiphenetole. In 1975, several groups investigated the structural features of mesogenic compounds in the crystalline state. Bryan et al. reported the crystal structures of p-n-butoxybenzoic acid [6] and p-methoxycinnamic acid [7]. Cortrait et al. described the crystal structures of the nematogenic propyl 4-(4′-methoxy-benzylidene-amino)-α-methylcinnamate [8] and the smectogenic (methyl-2-butyl)-2-(oxo-1-pentyl)-7 dihydro-9,10-phenanthrene [9]. Lesser et al. presented the crystal structure of the nematogenic compound 2,2′-dibromo-4,4′-bis-(p-methoxy-benzylideneamino)biphenyl [10]. Since then, a large number of crystal structures of liquid crystals have been determined and discussed in relation to the mesogenic behaviour.

In 1978, Bryan [11] reported on crystal structure precursors of liquid crystalline phases and their implications for the molecular arrangement in the mesophase. In this work he presented classical nematogenic precursors, where the molecules in the crystalline state form imbricated packing, and non-classical ones with cross-sheet structures. The crystalline-nematic phase transition was called displacive. The displacive type of transition involves comparatively limited displacements of the molecules from the positions which they occupy with respect to their nearest neighbours in the crystal. In most cases, smectic precursors form layered structures. The crystalline-smectic phase transition was called reconstitutive because the molecular arrangement in the crystalline state must alter in a more pronounced fashion in order to achieve the mesophase arrangement [12].

To date, the crystal structures of more than 200 mesogenic compounds are known. In this review, we wish to present a general overview of the crystal structures of mesogenic compounds up to the end of 1997. Unfortunately, it is not possible to consider the crystal structure determinations of carbohydrate liquid crystals [13, 14], metallomesogens [15–18], phasmid and biforked mesogens [19–22], perfluorinated mesogenic compounds [23–27], benzoic acids [6, 28–31], cinnamic acids [7, 32, 33], dicarboxylic acids [34, 35], cinnamate compounds [8, 36–40], and discotic liquid crystals [41–43] due to the lack of space.

2
Liquid Crystals Containing Two Attached Rings

2.1
Liquid Crystals Containing Two Aromatic Rings

2.1.1
4'-n-Alkyl-4-cyanobiphenyls (CBn)

The first report on the liquid crystalline properties of these compounds was published by Gray and Mosley [44] in 1976. The series of 4'-*n*-alkyl-4-cyanobiphenyls (CB*n*) have been widely studied by different methods due to their readily accessible nematic ranges around room temperature. The compounds have the phase sequences: crystal-nematic-isotropic for CB5, CB*10*, and monotropic nematic for CB3, CB4; crystal-smectic A-nematic-isotropic for CB9; crystal-smectic A-isotropic for CB*11*. The lower homologous CB2 is nonmesogenic. The general chemical structure of the compounds CB*n* is presented in Fig. 1.

The crystal structures of the mesogenic compounds CB*n* were reported by different authors [45–51]. Selected crystallographic and molecular data of the investigated compounds CB*n* are summarised in Table 1. Only the crystal structures of some compounds will be discussed here.

The mesogenic compound CB3 exhibits a solid state polymorphism (two solid phases: I and II). With regard to the packing, the crystal structure [46] of the higher melting modification I (melting point 338.4 K) shows that the cyano group of one molecule overlaps the phenyl ring adjacent to the cyano group of the neighbouring molecule, related by a centre of symmetry. The averaged N_{cyano}-C'_{phenyl} distance is 4.37 Å and the analogous averaged C_{cyano}-C'_{phenyl} distance is 3.98 Å. The crystal structure of the second solid phase II of CB3 remains unresolved.

The nonmesogenic compound CB2 is described here, because it shows a reversible distortive solid-solid phase transition at 290.8 K (transition enthalpy 0.9 kJ/mol) from the centrosymmetric low temperature phase I to the noncentrosymmetric high temperature phase II. The crystal structures of both solid phases I and II are very similar [45] as demonstrated in Fig. 2. The molecules are arranged in layers. The distances between the cyano groups of adjacent molecules are 3.50 Å N_{cyano}-N'_{cyano} and 3.35 Å N_{cyano}-C'_{cyano} for phase I and 3.55 Å N_{cyano}-N'_{cyano} and 3.43 Å N_{cyano}-C'_{cyano} for phase II. In the two

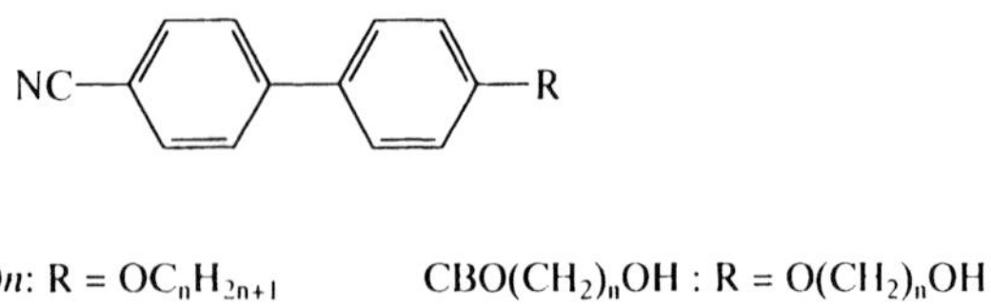

CB*n*: R = C$_n$H$_{2n+1}$ CBO*n*: R = OC$_n$H$_{2n+1}$ CBO(CH$_2$)$_n$OH : R = O(CH$_2$)$_n$OH

Fig. 1. Chemical structure of mesogenic cyanobiphenyls

Table 1. Crystal and molecular data of mesogenic 4'-n-alkyl-4-cyanobiphenyls (CBn)

Compound	Ref.	Crystal Phase	Space group	Z	Dihedral angle between the phenyl rings[c]/°	Selected intermolecular distances [Å]
CB2[a]	[45]	I	P2$_1$/c	4	0.7	3.50 N$_{cyano}$-N'$_{cyano}$ 3.35 N$_{cyano}$-C'$_{cyano}$
		II	P2$_1$	2	1.5	3.55 N$_{cyano}$-N'$_{cyano}$ 3.43 N$_{cyano}$-C$_{cyano}$
CB3	[46]	I	P2$_1$/c	4	42.8	4.37 N$_{cyano}$-C$_{phenyl}$[d] 3.98 C$_{cyano}$-C$_{phenyl}$[d]
CB4	[47]		P2$_1$/c	4	40.3	3.50 N$_{cyano}$-N'$_{cyano}$
CB5	[48]		P2$_1$/a	4	26.3	3.45 N$_{cyano}$-C'$_{cyano}$ antiparallel: 3.55 C$_{cyano}$-C'$_{phenyl}$[e] parallel: 3.77 C$_{cyano}$-C'$_{phenyl}$[e]
CB9[b]	[49]		P1̄	4	molecule 1: 36.0 molecule 2: 29.8	
CB10[b]	[50]		P2$_1$/n	4	2.2	
CB11[b]	[51]		P1̄	4	molecule 1: 30.2 molecule 2: 35.8	

In all the tables: Z is the number of molecules per unit cell; ' symbolizes the related atom of the adjacent molecule.
[a] Nonmesogenic compound.
[b] No intermolecular distances are given by the authors.
[c] Dihedral angle between the phenyl rings of the biphenyl group.
[d] Averaged distances to the six carbon atoms of the phenyl ring.
[e] Shortest distance to one carbon atom of the phenyl ring.

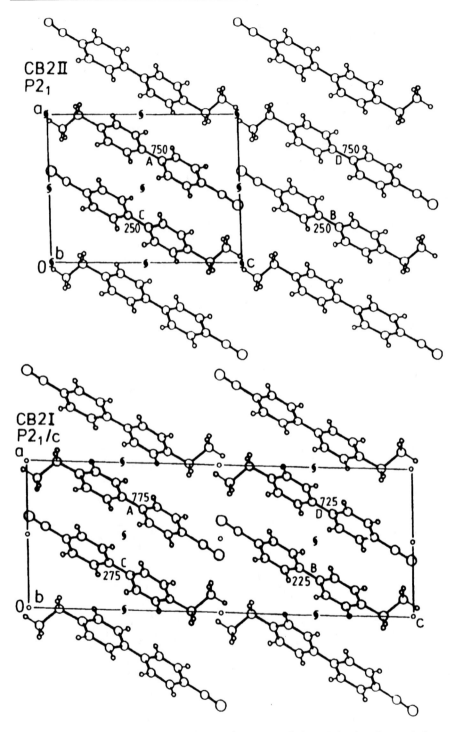

Fig. 2. Crystal structures of both solid phases of CB2, crystal phase II (*top*) and crystal phase I (*bottom*), along [010]. (Reprinted from [45])

solid state phases of CB2, the aromatic rings of the biphenyl group are approximately coplanar with negligible dihedral (twist) angles of 0.7° and 1.5°, due to packing effects in the lattice.

In the case of compound CB5 no solid state polymorphism was observed. The molecular packing in the crystalline state suggests an alternating arrangement of bilayers, with two parallel oriented molecules within one layer. The orientation of two neighbouring bilayers appears to be antiparallel [48]. In the case of antiparallel oriented molecules, the shortest intermolecular distance between the cyano group and one carbon atom of the phenyl ring attached to the pentyl chain is 3.55 Å C_{cyano}-C'_{phenyl}. The corresponding value for parallel packed molecules is 3.77 Å C_{cyano}-C'_{phenyl}, considering the shortest distance to one carbon atom of the phenyl ring attached to the cyano group. In contrast to most analogous compounds where the aliphatic chain possessing all-trans conformation, the pentyl chain is oriented nearly perpendicular to the plane of the attached aromatic ring (–90.5°).

The nematic phase of all the compounds CBn is characterized by a coherence length of about 1.4 times the elongated structure of the molecule. Based on this behaviour local associations in form of dimers with cyano-phenyl interactions were postulated. For the smectic A phase a partial bilayer arrangement of the molecules (S_{Ad}) is most likely. But there are also example for the smectic A phase with a monolayer (S_{A1}) or a bilayer (S_{A2}) arrangement of the molecules as well as a commensurate structure $S_{\tilde{A}}$. A large number of X-ray measurements were carried out in the liquid crystalline state to clear up the structural richness and variability (see Chap. 2, this Vol.; [52]).

2.1.2
4'-n-Alkoxy-4-cyanobiphenyls (CBOn)

In this section, we will describe the crystal structures of the mesogenic 4'-n-alkoxy-4-cyano-biphenyls (Fig. 1) [53–57]. The compounds have the phase sequences: crystal-nematic-isotropic for CBO5--CBO7 and monotropic nematic for CBO1–CBO4; crystal-smectic A-nematic-isotropic for CBO8. The crystal and molecular data of the investigated compounds CBOn are summarised in Table 2.

Walz et al. [53] reported on the crystal and molecular structures of four mesogenic compounds of the series 4'-alkoxy-4-cyanobiphenyl (CBO1–CBO4). Compound CBO4 crystallises in the noncentrosymmentric space group Pca2₁. The two crystallographically independent molecules are arranged with their longest extension along the c-axis. The crystal packing in CBO4 (view along [100]) is shown in Fig. 3. The cyano dipoles have an almost perfect parallel arrangement. With regard to intermolecular interactions, no contacts between cyano groups or cyano groups with phenyl rings were observed. Pyroelectric, piezoelectric, and optical second harmonic generation (SHG) measurements in the solid state of CBO4 were carried out [58]. Ferroelectric switching in the solid state of a mesogenic compound has been observed for the first time in

Table 2. Crystal and molecular data of mesogenic 4'-n-alkoxy-4-cyanobiphenyls (CBOn)

Compound	Ref.	Crystal Phase	Space group	Z	Dihedral angle between the phenyl rings[a]/°	Selected intermolecular distances [Å]
CBO1	[53]		$P2_1/n$	8	molecule 1: 44.1 molecule 2: 43.6	4.41 N_{cyano}-C'_{phenyl}[b] 4.17 C_{cyano}-C'_{phenyl}[b]
CBO2	[53]		$C2/c$	16	molecule 1: 45.9 molecule 2: 33.4	4.17 N_{cyano}-C'_{phenyl}[b] 3.96 C_{cyano}-C'_{phenyl}[b]
CBO3	[53]		$P2_1/c$	4	47.4	3.50 N_{cyano}-N_{cyano} 3.40 N_{cyano}-C_{cyano}
CBO4	[53]		$Pca2_1$	8	molecule 1: 29.9 molecule 2: 39.0	
CBO5	[54]		$P2_1/n$	4	0.8	3.40 N_{cyano}-N'_{cyano} 3.34 N_{cyano}-C_{cyano}
CBO6	[55]		$P2_1/a$	8	molecule 1: 36.0 molecule 2: 26.0	3.57 $N1_{cyano}$-$N2'_{cyano}$ 3.40 $N1_{cyano}$-$C2'_{cyano}$ 3.56 $C1_{cyano}$-$N2'_{cyano}$
CBO7	[55]	needle	$P\bar{1}$	4	molecule 1: 37.0 molecule 2: 40.0	3.52 $N1_{cyano}$-$N2_{cyano}$ 3.39 $N1_{cyano}$-$C2'_{cyano}$ 3.55 $C1_{cyano}$-$N2'_{cyano}$
CBO8	[56] [57]	square-plate needle long parallelepiped	$C2/c$ $P\bar{1}$ $P2_1/a$	8 4 4	31.1 molecule 1: 38.2 molecule 2: 38.9 2.2	3.41–3.64 N_{cyano}-C'_{cyano} 3.54 $N1_{cyano}$-$N2'_{cyano}$ 3.40 $N2_{cyano}$-$C1'_{cyano}$ 3.39 N_{cyano}-C'_{phenyl}[c] 3.69 N_{cyano}-C'_{phenyl}[c]

[a] Dihedral angle between the phenyl rings of the biphenyl group.
[b] Averaged distances to the six carbon atoms of the phenyl ring attached to the cyano group.
[c] Distances between the cyano group and two carbon atoms of the phenyl rings.

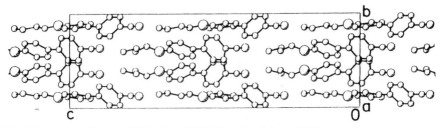

Fig. 3. Crystal packing in CBO4 along [100] (Reprinted from [53])

CBO4. The spontaneous polarisation was measured by the pulse pyroelectric technique (P_S = 46 nC/cm^2). The piezoelectric coefficient evaluated for CBO4 was d_{31} = 1.6 pC/N. The estimation of the efficiency of the second harmonic generation for compound CBO4 gives the value three times more than for quartz.

For the compound CBO7 four different solid phases were found. The crystal structures of the needle and the square plate crystals were described by Hori et al. [55, 56]. The needle crystal contains two crystallographically independent molecules. The close contacts of the cyano groups of the molecules 1 and 2 form a dimeric arrangement with cyano-cyano distances of 3.52 Å $N1_{cyano}$-$N2'_{cyano}$ and 3.39 Å $N1_{cyano}$-$C2'_{cyano}$ which is further paired with the neighbouring dimer to form a tetramer (cyano-cyano distances of 3.57 Å). The molecules are arranged in layers. The plate crystals have a distinct smectic-like structure composed of bilayers. The cyano groups are closely arranged between antiparallel molecules within a bilayer, resulting in infinite networks with N_{cyano}-C'_{cyano} distances of 3.41–3.64 Å.

Four crystalline phases were also found for CBO8, and the crystal structures of the needle and the long parallelepiped crystals were determined by Hori et al. [57]. The molecular packing in the needle crystals of CBO8 is almost the same as that found in the needle crystals of CBO7 [55]. In the long parallelepiped crystal, the cyano group is not close to another cyano group, but is directed towards the biphenyl group of the adjacent molecule with contact distances of 3.39 Å N_{cyano}-C'_{phenyl} (phenyl ring attached to the cyano group) and 3.69 Å N_{cyano}-C'_{phenyl} (phenyl ring attached to the octyloxy group), respectively. The dihedral angle between the aromatic rings of the biphenyl group is 2.2°.

All the investigated CBOn compounds show the formation of dimers in the liquid crystalline phases as previously described for the CBn series (see Sect. 2.1.1).

2.1.3
4'-(Hydroxy-1-n-alkoxy)-4-cyanobiphenyls (CB(CH$_2$)$_n$OH) and Derivatives

The crystal structure of the mesogenic compound 4'-(4-hydroxy-1-n-butoxy)-4-cyanobiphenyl (CBO(CH$_2$)$_4$OH) was described by Gehring et al. [59]. In

1993, Zugenmaier and Heiske [60] presented the crystal and molecular structures of the homologous series of 4'-(hydroxy-1-n-alkoxy)-4-cyanobiphenyls (CBO(CH$_2$)$_n$OH) n = 4, 5, 7–11). The chemical structure of the compounds is shown in Fig. 1. All compounds of the series exhibit a nematic phase. The crystal and molecular data of the investigated compounds CBO(CH$_2$)$_n$OH and some derivatives are presented in Table 3.

All compounds in the series are packed in an antiparallel fashion with more or less strong intermolecular interactions between the polar end groups. In the case of compound CBO(CH$_2$)$_4$OH, no contacts between cyano groups or cyano groups with phenyl rings can be observed, but hydrogen bonds between terminal hydroxy and cyano groups exist (distance N$_{cyano}$-O'$_{hydroxy}$ = 2.89 Å) [59, 60]. The compounds CBO(CH$_2$)$_n$OH with n = 7–11 have isomorphous layer structures [60]. The molecular packing of CBO(CH$_2$)$_7$OH is demonstrated in Fig. 4. The cyano groups form layers parallel to the ab plane. Within such layers, every cyano group is surrounded by four adjacent cyano groups, indicating dipole-dipole interactions. Hydrogen bonds between the hydroxy groups are formed (intermolecular hydrogen bond lengths of 2.68–2.83 Å), but between the cyano and hydroxy groups no hydrogen bonds exist.

For compound 4'-(2-hydroxymethyl-3-hydroxy-propyloxy)-4-cyanobiphenyl (CBOCH$_2$CH(CH$_2$OH)$_2$) two types of hydrogen bonds were found, one between the cyano and hydroxy groups (distance N$_{cyano}$-O$_{hydroxy}$ = 2.95 Å), and the other among the hydroxy groups (distance O$_{hydroxy}$-O'$_{hydroxy}$ = 2.77 Å). The contact distance between the cyano groups of adjacent molecules is around 3.58 Å [60].

The crystal and molecular structures of the chiral (R)-4'-(2,3-dihydroxypropyloxy)-4-cyanobiphenyl (CBOCH$_2$(HCOH)H$_2$COH) and the achiral 4'-(cis-cis-3,5-dihydroxy-cyclohexyloxy)-4-cyanobiphenyl (CBOC$_6$H$_9$(OH)$_2$) were determined by Joachimi et al. [61]. The first compound exhibits a cholesteric phase [62] and the second one forms a monotropic nematic phase [61]. The crystal packing of the chiral compound occurs in sheets with a dense hydrogen bonding network between adjacent sheets, whereby the cyano groups are incorporated into the hydrogen bonding networks of the diol groups (intermolecular hydrogen bond lengths: O$_{hydroxy}$-N'$_{cyano}$ = 2.91 Å; O$_{hydroxy}$-O'$_{hydroxy}$ = 2.69 Å). The molecules are arranged antiparallel within the layers,

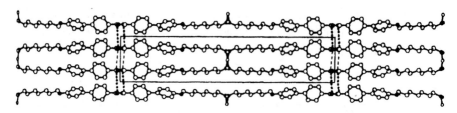

Fig. 4. Molecular packing of HO(CH$_2$)$_7$OCB as projection on the bc plane. *Dashed lines* represent stronger dipole-dipole interactions and *dotted lines* weaker contacts. (Reprinted from [60])

Table 3. Crystal and molecular data of mesogenic 4'-(hydroxy-1-n-alkoxy)-4-cyanobiphenyls (CBO(CH$_2$)$_n$OH) and some derivatives

Compound	Ref.	Space group	Z	Dihedral angle between the phenyl rings[a]/°	Selected intermolecular distances [Å]
CBO(CH$_2$)$_4$OH	[59,60]	P$\bar{1}$	2	22.3	2.89 N$_{cyano}$-O$'_{hydroxy}$
CBO(CH$_2$)$_5$OH	[60]	C2/c	8	37.7	2.96 N$_{cyano}$-O$'_{hydroxy}$ 3.75 N$_{cyano}$-C$'_{cyano}$
CBO(CH$_2$)$_7$OH	[60]	P2/a	4	31.4	3.56 N$_{cyano}$-C$'_{cyano}$
CBO(CH$_2$)$_8$OH	[60]	P2/a	4	30.8	3.50 N$_{cyano}$-C$'_{cyano}$
CBO(CH$_2$)$_9$OH	[60]	P2/a	4	31.1	3.54 N$_{cyano}$-C$'_{cyano}$
CBO(CH$_2$)$_{10}$OH	[60]	P2/a	4	30.9	3.43 N$_{cyano}$-C$'_{cyano}$
CBO(CH$_2$)$_{11}$OH	[60]	P2/a	4	31.1	3.57 N$_{cyano}$-C$'_{cyano}$
CBOCH$_2$CH(CH$_2$OH)$_2$	[60]	P2$_1$/n	4	36.9	2.95 N$_{cyano}$-O$'_{hydroxy}$ 2.77 O$_{hydroxy}$-O$'_{hydroxy}$ 3.58 N$_{cyano}$-C$'_{cyano}$
CBOCH$_2$(HCOH)CH$_2$OH	[61]	P2$_1$	2	13.0	2.91 N$_{cyano}$-O$'_{hydroxy}$ 2.69 O$_{hydroxy}$-O$'_{hydroxy}$
CBOC$_6$H$_9$(OH)$_2$	[61]	P2$_1$/c	4	23.0	2.96 N$_{cyano}$-O$_{water}$
CBO(CH$_2$)$_7$OBC	[63]	P2$_1$/n	4	39.0	2.73 O$_{hydroxy}$-O$'_{hydroxy}$

[a] Dihedral angle between the phenyl rings of the biphenyl group.

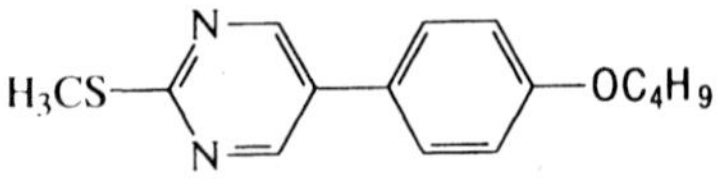

Fig. 5. Chemical structure of 2-methylthio-5-(4′-n-butyloxyphenyl)pyrimidine

resulting in a "pseudo-head-to-tail-arrangement" of the molecules in neighbouring layers. This kind of packing also prevails in compound CBO-$C_6H_9(OH)_2$. The hydrogen bonding scheme of this compound involves a water molecule (intermolecular hydrogen bond lengths: O_{water}-$O'_{hydroxy}$ = 2.71 Å; N'_{cyano}-O_{water}- = 2.96 Å; $O_{hydroxy}$-$O'_{hydroxy}$ = 2.73 Å).

In 1991, Malpezzi et al. [63] described the crystal structure of α,ω-bis(4-cyanobiphenyl-4′-oxy)heptane (CBO(CH$_2$)$_7$OBC). The flexible spacer chain (CH$_2$)$_7$ shows an all-trans conformation. The dihedral angle between the aromatic rings of the biphenyl group is 39°. They found that the long axes of the mesogenic cyanobiphenyl groups are coplanar with the chain plane and are tilted at a mean angle of 17.4° relative to the spacer chain axis, giving an angle between the mesogenic units of 34°. With regard to intermolecular interactions, no contacts between cyano groups or cyano groups with phenyl rings can be observed.

2.1.4
2-Methylthio-5-(4′-n-butyloxyphenyl)-pyrimidine

Hartung and Rapthel [64] described the crystal structure of the mesogenic 2-methylthio-5-(4′-n-butyloxyphenyl)-pyrimidine which forms a monotropic smectic A phase. The chemical structure of this compound is presented in Fig. 5. The compound crystallises in the triclinic space group P$\bar{1}$ with two molecules per unit cell. The molecules adopt a fully stretched and nearly planar form. The pyrimidine ring is nearly planar. The dihedral angle between the phenyl and the pyrimidine rings is 22.7°. The molecules are arranged parallel to each other.

2.2
Liquid Crystals Containing One Aromatic and One Aliphatic Ring

2.2.1
trans-4-n-Alkyl-(4′-cyanophenyl)-cyclohexanes (PCHn)

The synthesis of the mesogenic trans-4-n-alkyl-(4′-cyanophenyl)-cyclohexanes (PCHn) was described by Eidenschink et al. [65] in 1977. Most of the compounds exhibit a nematic phase close to room temperature. The chemical structure of the mesogenic PCHn is shown in Fig. 6. During the past few years, the crystal structures of some mesogenic phenylcyclohexanes were published [66–70]. Selected crystallographic and molecular data of the investigated compounds PCHn are presented in Table 4.

Table 4. Crystal and molecular data of mesogenic *trans*-4-*n*-alkyl-(4'-cyanophenyl)-cyclohexanes (PCH*n*)

Compound	Ref.	Crystal Phase	Space group	Z	Dihedral angle between the phenyl and cyclohexyl groups[a]/°	Selected intermolecular distances [Å]
PCH3	[67]	I	C2/c	8	81.5	3.57 N_{cyano}–N'_{cyano} 3.38 N_{cyano}–C'_{cyano}
PCH4	[68]	I	P2$_1$/c	4	68.6	4.80 N_{cyano}–C'_{phenyl}[b] 4.54 C_{cyano}–C'_{phenyl}[b]
PCH8	[69]	A	P2$_1$/c	4	64.9	3.48 N_{cyano}–N'_{cyano} 3.86 N_{cyano}–N'_{cyano} 3.45 N_{cyano}–C'_{cyano} 3.55 N_{cyano}–C'_{cyano}
	[70]	B	P$\bar{1}$	2	36.5	4.51 N_{cyano}–C'_{phenyl}[b] 4.31 C_{cyano}–C'_{phenyl}
PCH9	[70]	A	P2$_1$/c	4	65.9	3.45 $N1_{cyano}$–$N1'_{cyano}$ 3.47 $N1_{cyano}$–$C1'_{cyano}$
	[70]	B	P$\bar{1}$	4	molecule 1: 45.9 molecule 2: 41.3	3.56 $N1_{cyano}$–$N2'_{cyano}$ 3.43 $N1_{cyano}$–$C2'_{cyano}$ 3.54 $N2_{cyano}$–$C1'_{cyano}$

[a] For the calculation of the best planes of the cyclohexyl ring only the four central carbon atoms were used.
[b] Averaged distances to the six carbon atoms of the phenyl ring.

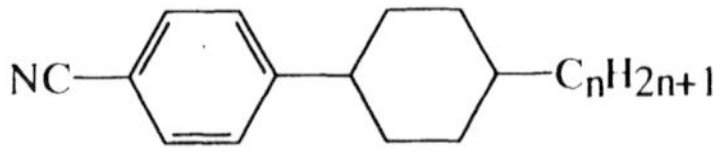

Fig. 6. Chemical structure of mesogenic trans-4-*n*-alkyl-(4′-cyanophenyl)-cyclohexanes (PCH*n*)

Compound PCH3 shows a solid state polymorphism [66] (three solid phases: I, II, and III). The crystal structure of the higher melting modification I (melting point 316 K and melting enthalpy 19 kJ/mol) of PCH3 was described by Foitzik et al. [67]. They found that the two rings of the molecule lie nearly perpendicular (81.5°) to each other. No contacts between cyano groups or cyano groups with phenyl rings can be observed. Unfortunately, the crystal structures of the two other solid phases II (melting point 308.6 K and melting enthalpy 12.8 kJ/mol) and III (melting point 294 K and melting enthalpy 13.5 kJ/mol) of PCH3 are still unknown.

In the case of compound PCH4 no solid state polymorphism was observed. The molecules are extended parallel to each other and lie perpendicular to [010] alternating in a head-to-tail arrangement [68]. The dihedral angle between the cyclohexyl and the phenyl groups is 68.6°. Each cyano group of one molecule overlaps with the cyano group of the neighbouring molecule. The intermolecular distances between the cyano groups of adjacent molecules are 3.57 Å N_{cyano}-N'_{cyano} and 3.38 Å N_{cyano}-C'_{cyano}, respectively.

For compound PCH8 two different solid phases were found. The crystal structures of the two modifications A and B of PCH8 were described [69, 70]. In case of the modification A of PCH8 [69], the molecules are arranged in a head-to-tail manner parallel to each other with an overlapping of the octylcyclohexyl groups of the molecules. Additionally, each cyanophenyl group overlaps with the cyanophenyl group of the neighbouring molecule related by a centre of symmetry. The averaged N_{cyano}-C'_{phenyl} distance is 4.80 Å and the analogous C_{cyano}-C'_{phenyl} distance is 4.54 Å. The dihedral angle between the cyclohexyl and the phenyl groups is 64.9°. In the modification B of PCH8 [70], the molecules are arranged along the direction [211] with infinite chains of overlapped cyano groups along [100]. The contact distances between the cyano groups of adjacent molecules are presented in Table 4. The dihedral angle between the cyclohexyl and the phenyl groups is 36.5°.

Two different solid phases were also found for compound PCH9 [70]. The crystal structures of the two modifications A and B of PCH9 are shown in Fig. 7. In the case of the modification A, the molecular packing is isostructural to the modification A of PCH8 [69]. By contrast, the crystal structure of the modification B of PCH9 contains two crystallographically independent molecules which are arranged parallel to [011]. The close contacts of the cyano groups of the molecules 1 and 2 form a dimeric arrangement with cyano distances of 3.56 Å $N1_{cyano}$-$N2'_{cyano}$ and 3.43 Å $N1_{cyano}$-$C2'_{cyano}$. Likewise each molecule 1 forms with the adjacent molecule 1′ a dimer related by a centre of symmetry with antiparallel overlapping of the cyano groups at distances of

3.47 Å. As a result of this, a clustering occurs of four molecules through cyano-cyano contacts resulting in an island.

Although all the investigated PCH*n* compounds crystallise in the all-trans conformation, the twist angles between the phenyl and the cyclohexyl rings vary (Table 4) due to the different packing of the molecules in the crystalline state. It is interesting that the melting temperatures and enthalpies of the two

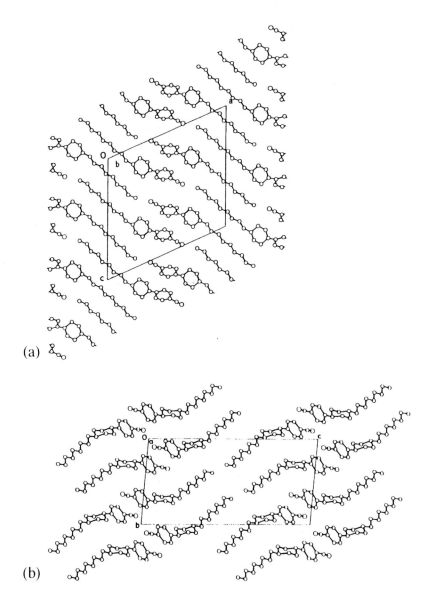

(a)

(b)

Fig. 7a,b. Crystal structures of the two modifications A and B of PCH9: **a** crystal packing in PCH9A along [010]; **b** crystal packing in PCH9B along [100]. (Reprinted from [70])

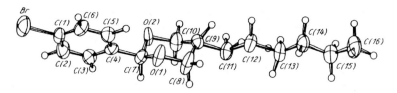

Fig. 8. Molecular structure of 5-hexyl-2-(4′-bromophenyl)-1,3-dioxane (Reprinted from [71])

modifications A of PCH8 and PCH9 are higher compared with the modifications B of PCH8 and PCH9, respectively. This indicates that in this particular case the modifications with cyano-phenyl overlapping are more stable than the modifications with cyano-cyano overlapping. The investigated compounds form a nematic mesophase with a correlation length typical for molecular associations. For instance, in the nematic phase of PCH8 the correlation length was found to be 32.1 Å, but the extended molecular length is 21.0 Å [69].

2.2.2
5-Hexyl-2-(4′-bromophenyl)-1,3-dioxane and 5-Heptyl-2-(4′-cyanophenyl)-1,3-dioxane

Hartung and co-workers [71] investigated the crystal structure of the mesogenic 5-hexyl-2-(4′-bromophenyl)-1,3-dioxane (HBPD) which shows a monotropic smectic A mesophase. This compound crystallises in the monoclinic space group C2/c with eight molecules per unit cell. The molecular structure of the compound is presented in Fig. 8. The molecules adopt a fully stretched form, and both the alkyl and phenyl groups are in equatorial positions with respect to the chair conformation of the dioxane ring. The molecules show a nearly perfect parallel arrangement in the crystal lattice.

Bellad et al. [72] presented the crystal structure of the 5-heptyl-2-(4′-cyanophenyl)-1,3-dioxane which exhibits a smectic phase. This compound crystallises in the orthorhombic space group $P2_12_12_1$ with four molecules per unit cell. The molecules are arranged in layers.

2.3
Liquid Crystals Containing Two Aliphatic Rings

2.3.1
trans,trans-4′-n-Alkylbicyclohexyl-4-carbonitriles (CCHn) and Derivatives

In 1977, Eidenschink et al. [65] reported on the new liquid crystalline trans, trans-4′-n-alkylbicyclohexyl-4-carbonitriles (cyclohexylcyclohexanes, CCHn). The general chemical structure of the mesogenic CCHn compounds is presented in Fig. 9.

Fig. 9. Chemical structure of mesogenic trans,trans-4′-alkylbicyclohexyl-4-carbonitriles (CCHn)

The crystal and molecular structures of three compounds of this homologous series (alkyl = propyl (CCH3), pentyl (CCH5), and heptyl (CCH7)) were determined by Haase and co-workers [73, 74]. All three compounds possess a nematic phase, whereas CCH3 and CCH5 exhibit additionally three monotropic smectic phases and two monotropic smectic phases, respectively. Brownsey and Leadbetter [75] have identified by X-ray investigations the smectic phases of CCH3 and CCH5, which have the highest transition points. They belong to the bilayered smectic B phase. The crystal and molecular data of the CCH*n* compounds and some derivatives are summarised in Table 5.

With regard to the molecular structure of CCH*n*, both cyclohexyl rings having a chair conformation are substituted in the equatorial positions and the alkyl chain is completely extended in the all-trans conformation. The cyclohexyl rings are nearly coplanar. The crystal structures of the investigated CCH*n* show that various types of molecular overlapping are present in the crystal. The molecular packing in the crystalline state is quite different in all three compounds.

CCH3 shows the formation of dimers related by a centre of symmetry and close contacts of the cyano groups between neighbouring molecules (see Table 5). The molecules are arranged in layers and overlap more or less with the bicyclohexyl units.

Compound CCH5 exhibits a solid state polymorphism (three solid phases: I, II, and III). Only the crystal structures of the phases I and II of CCH5 were described [73, 74]. The transition temperature from the phase II to the phase I is 332.4 K (transition enthalpy: 6.1 kJ/mol). The I-phase crystallises in a centrosymmetric monoclinic space group and the II-phase in a noncentrosymmetric orthorhombic one. The crystal structures of the two phases I and II of CCH5 are demonstrated in Fig. 10. Both solid phases have layered structures and within the layers the molecules are tilted with respect to the layer normal. In the I-phase, the central cyclohexyl rings of neighbouring molecules overlap in a head-to-tail manner forming a herringbone structure. The molecular packing of CCH7 is almost the same as that found in the II-phase of CCH5. The cyano groups form an infinite stacking around the 2_1-axis in CCH7 and CCH5 II. The contact distances between the cyano groups of adjacent molecules are slightly shorter in CCH7 compared with CCH5 II.

It is interesting that all three compounds show small angle reflections in the liquid crystalline state which indicates the formation of associates with a length of about twice the molecular length (for CCH5: 17.5 Å in the crystal phase I, 31.2 Å in the S_B phase, and 27.2 Å in the nematic phase) [73].

In 1991, Ji et al. [76] described the crystal and molecular structures of the two mesogenic 1-(4'-cyanocyclohexyl)-*trans*-4-(1-buten-4-yl)-cyclohexane (CCBCH) and 1-(4'-cyanocyclohexyl)-*trans*-4-(2-penten-5-yl)-cyclohexane (CC-PCH). The crystal structure of CCPCH was also investigated by Gupta et al. [77]. Both compounds possess a nematic phase, whereas CCPCH shows additionally a monotropic smectic A mesophase within a very small temperature range. The crystal packing of the compounds, the arrangement of the molecules in the layer and the orientation of the layer with respect to the crystal lattice are different. The crystal structure of CCBCH contains two crystallographically

Table 5. Crystal and molecular data of mesogenic *trans,trans*-4'-*n*-alkylbicyclohexyl-4-carbonitriles (CCH*n*) and some derivatives

Compound	Ref.	Crystal Phase	Space group	Z	Selected intermolecular distances [Å]
CCH3	[73]		$P\bar{1}$	2	3.55 N_{cyano}-N'_{cyano} 3.27 N_{cyano}-C'_{cyano}
CCH5	[73]	I	$P2_1/c$	4	
	[74]	II	$P2_12_12_1$	4	5.01 N_{cyano}-N'_{cyano}
CCH7	[73]		$P2_12_12_1$	4	4.01 N_{cyano}-N'_{cyano} 3.72 N_{cyano}-C'_{cyano}
CCBCH	[76]		$P2_1/c$	8	
CCPCH	[76]		$P2_1/n$	4	
	[77]				
PEBCH	[78]		$P2_12_12_1$	4	
CCCHC	[79]		$P2_12_12_1$	4	

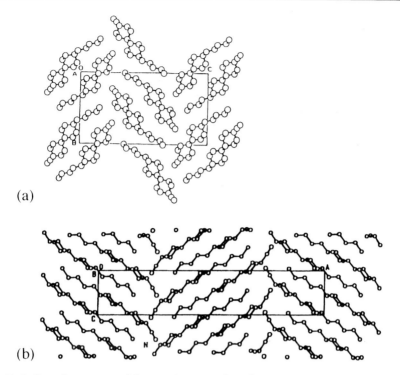

Fig. 10a,b. Crystal structures of the two phases I and II of CCH5: **a** crystal packing in CCH5 I along [100]; **b** crystal packing in CCH5 II along [010]. (Reprinted from [73, 74])

independent molecules. Compound CCBCH shows a parallel imbrication of molecules in the crystal, whereas compound CCPCH displays a herringbone relationship on molecular stacks in the crystal structure [76].

The crystal structure of the mesogenic 4-(3''-pentenyl)-4'-(ethoxy)-1,1'-bicyclohexane (PEBCH) was determined by Nath et al. [78]. This compound exhibits a nematic phase. The molecules are packed parallel to each other in an imbricated mode.

In 1989, Krieg et al. [79] presented the crystal structure of the smectogenic *trans*-4-(*cis*-4-cyano-cyclohexyl)cyclohexyl *trans*-4-*n*-heptylcyclohexanoate (CCCHC). In contrast to all other substituents which are equatorially bonded in 1,4-positions to the cyclohexyl rings, the cyano group is axially attached. The molecules adopt a fully stretched conformation. Compound CCCHC has a tilted layer structure with a herringbone arrangement of layers.

2.3.2
trans,trans-4,4'-m-n-Dialkyl-(1α, 1'-bicyclohexyl)-4β-carbonitriles (CCNnm)

In 1989, Walz et al. [80] reported on the crystal and molecular structures of four mesogenic compounds of the series *trans, trans*-4,4'-*m-n*-dialkyl-(1α,1'-bicyclohexyl)-4β-carbonitrile (CCNnm). The chemical structure of the mesogenic CCNnm is presented in Fig. 11.

Fig. 11. Chemical structure of mesogenic *trans,trans*-4,4'-dialkyl-(1,1'-bicyclohexyl)-4β-carbonitriles (CCN*nm*)

Table 6. Crystal data and phase sequences of mesogenic *trans,trans*-4,4'-*n*-*m*-dialkyl-(1,1'-bicyclohexyl)-4β-carbonitriles (CCN*nm*)

Compound	Ref.	Space group	Z	Dihedral angle between the cyclohexyl rings[a]/°
CCN33	[80]	Pbca	8	77.3
				8.4
CCN35	[80]	P2₁/c	4	81.6
CCN38	[80]	P2₁/c	4	81.9
CCN46	[80]	P2₁/c	4	7.3

[a] For the calculation of the best plane of the cyclohexyl ring only the four central carbon atoms were used.

The crystal structures of the following four compounds CCN33, CCN35, CCN38, and CCN46 were determined. The compounds have different phase sequences: C-[N]-I for CCN33; C-N-I for CCN35; C-[SmA]-N-I for CCN38; and C-[SmB]-N-I for CCN46. The crystal and molecular data of the investigated CCN*nm* are summarised in Table 6. In all four structures the cyano groups occupy axial positions whereas the alkyl groups occupy equatorial ones. For the calculation of the best planes of the cyclohexyl rings only the four central carbon atoms were used. The cyclohexyl rings in CCN35 and CCN38 are approximately perpendicular to each other with dihedral angles between the cyclohexyl rings of 81.6° (CCN35) and 81.9° (CCN38), respectively. In the case of compound CCN46, the cyclohexyl rings are nearly coplanar with a dihedral angle of 7.3°. Compound CCN33 crystallises in the noncentrosymmetric space group Pbca with a statistical mixture of molecules with coplanar and twisted arrangements of the cyclohexyl rings. With respect to the crystal packing of these compounds, no contacts between the polar cyano groups can be observed.

3
Liquid Crystals Containing Three Attached Rings

3.1
Liquid Crystals Containing Three Aromatic Rings

3.1.1
4-*Cyano*-4'-*n*-pentyl-*p*-terphenyl

To our knowledge, the crystal structure of only one mesogenic terphenyl compound has been described in the literature up to now [81]. The chemical

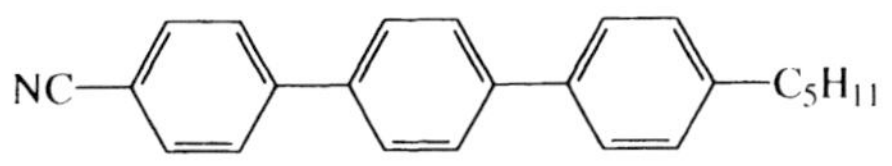

Fig. 12. Chemical structure of the mesogenic 4-cyano-4'-*n*-pentyl-*p*-terphenyl (T15)

structure of 4-cyano-4'-*n*-pentyl-*p*-terphenyl (T15) is shown in Fig. 12. This compound, first prepared by Gray et al. [82], exhibits a nematic phase with a wide temperature range.

Compound T15 shows solid state polymorphism (three solid phases: I, II, and III). Only the crystal structure of the low temperature solid phase I could be solved. This phase is monoclinic, space group P2₁/c, with four molecules in the unit cell. The dihedral angles between neighbouring phenyl groups are 40.4 and 29.5°, respectively. The corresponding angle between the two terminal phenyl rings is 69.3°. The molecules are extended parallel to each other and lie perpendicular to [010] alternating in a head-to-tail configuration. There are no intermolecular interactions between cyano groups or cyano groups with phenyl rings. For the two higher temperature phases II and III only the lattice parameters and the crystal systems could be determined by X-ray powder analyses [81]. X-ray investigations in the nematic phase of T15 indicated that dimeric as well as monomeric units exist.

3.1.2
5-Phenyl-2-(4'-n-butoxyphenyl)-pyrimidine
and 2-Phenyl-5-(4'-n-pentoxyphenyl)-pyrimidine

In this section we will report on the crystal structure analyses of mesogenic 2,5-diphenyl pyrimidines. The crystal structure of 5-phenyl-2-(4'-*n*-butoxy-phenyl)-pyrimidine (5-PBuPP) and 2-phenyl-5-(4'-*n*-pentoxyphenyl)-pyrimi-dine (2-PPePP) were determined by Winter et al. [83, 84]. Compound 5-PBuPP forms a monotropic nematic phase, whereas compound 2-PPePP exhibits a smectic A mesophase within a wide temperature range. The chemical structure of the mesogenic 2,5-diphenyl pyrimidines is shown in Fig. 13.

The crystal and molecular data of the investigated 2,5-diphenyl pyrimidines are presented in Table 7. The molecules of both compounds adopt a stretched form. In the case of compound 2-PPePP, the pyrimidine ring is approximately coplanar with the phenyl rings attached to it in the 2- and 5-positions. By contrast, the phenyl rings are twisted with respect to the central pyrimidine ring in 5-PBuPP.

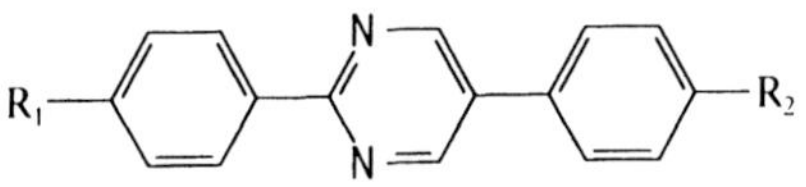

Fig. 13. Chemical structure of mesogenic 2,5-diphenyl pyrimidines

Table 7. Crystal and molecular data of mesogenic 2,5-diphenyl pyrimidines

Compound	Ref.	R_1	R_2	Space group	Z	Dihedral angle between 2-phenyl and pyrimidine rings/°	Dihedral angle between 5-phenyl and pyrimidine rings/°
5-PBuPP	[83]	H_9C_4O-	$-H$	$P2_1/c$	4	7.6	37.7
2-PBePP	[84]	$H-$	$-OC_5H_{11}$	$P\bar{1}$	4	8.0	8.0

Compound 5-PBuPP is isostructural with the nonmesogenic 5-PPrPP. With regard to the crystal packing of these compounds, the molecules are arranged nearly parallel to each other forming sheets parallel to the crystallographic *ac*-plane [83]. The crystal structure of 2-PPePP contains two crystallographically independent molecules. The molecules are arranged exactly parallel to each other forming sheets as well as a tilted layer structure [84].

3.2
Liquid Crystals Containing Two Aromatic and One Aliphatic Rings

3.2.1
4-n-Alkyl-4′-(4″-m-alkylcyclohexyl)-biphenyls (BCH) and Derivatives

The liquid crystalline biphenylcyclohexanes (BCH) were first described by Eidenschink et al. [85, 86]. The chemical structure of the mesogenic biphenyl-cyclohexanes is presented in Fig. 14.

The crystal structures of the mesogenic biphenylcyclohexanes BCH*5CN*, BCH*30*, BCH*52*, and BCH*52F* have been determined [87, 88]. The compounds show different phase sequences: C-N-I for BCH*5CN* and BCH*52F*; C-SmB-I for BCH*30*; and C-SmB-N-I for BCH*52*. The crystallographic data and some molecular features of BCH are summarised in Table 8.

In the case of BCH*5CN* and BCH*52*, the aromatic rings of the biphenyl group are approximately coplanar with dihedral angles of 2.4 and 3.9°, respectively. By contrast, the biphenyl group is twisted by 23.8° in BCH*30* and 41.9° in BCH*52F*. The dihedral angles between the phenyl and the cyclohexyl rings are 81.7° (BCH*5CN*), 75.8° (BCH*30*), 89.7° (BCH*52*), and 48.7° (BCH*52F*) (see Table 8). The molecules of BCH*5CN* [87] are arranged in a head-to-tail manner parallel to each other, lying perpendicular to [010]. The alternating head-to-tail arrangement results in a stack formation along the *b*-axis with infinite stacking of the cyano groups along [010]. The contact distance between the cyano groups of adjacent molecules is around 3.5 Å. In the case of BCH*30* and BCH*52*, the molecules lie with their long axes nearly parallel in a head-to-tail arrangement. The molecules in BCH*52F* lie perpendicular to [100] and are inclined at an angle of 40° to each other in the projection onto the (100) plane. There is no contact between the fluorine atoms of neighbouring molecules [88].

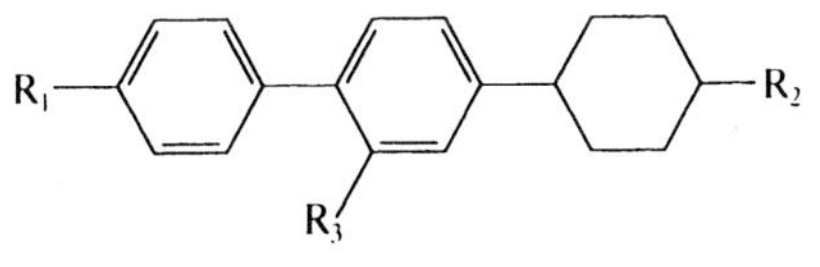

Fig. 14. Chemical structure of mesogenic biphenylcyclohexanes (BCH)

Table 8. Crystal and molecular data of mesogenic biphenylcyclohexanes (BCH)

Compound	Ref.	R_1	R_2	R_3	Space group	Z	Dihedral angle between the phenyl rings/°	Dihedral angle between the central phenyl and the cyclohexyl rings/°
BCH5CN	[87]	NC-	-C_5H_{11}	-H	$P2_1/c$	4	2.4	81.7
BCH30	[87]	H-	-C_3H_7	-H	$P2_1/c$	4	23.8	75.8
BCH52	[88]	H_5C_2-	-C_5H_{11}	-H	$C2/c$	8	3.9	89.7
BCH52F	[88]	H_5C_2-	-C_5H_{11}	-F	Pbcn	8	41.9	48.7

For the calculation of the best planes of the cyclohexyl ring only the four central carbon atoms were used.

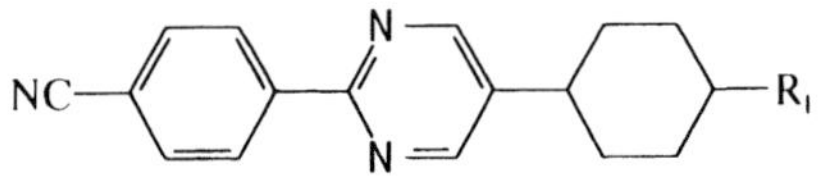

Fig. 15. Chemical structure of mesogenic 5-(4'-alkylcyclohexyl)-2-(4''-cyanophenyl)-pyrimidines

3.2.2
5-(4'-n-Alkylcyclohexyl)-2-(4''-cyanophenyl)-pyrimidines

Mandal et al. [89–91] investigated the crystal structures of three members of the homologous series of 5-(4'-*n*-alkylcyclohexyl)-2-(4''-cyanophenyl)-pyrimidines. The crystal structures of the ethyl (ECCPP), pentyl (PCCPP), and heptyl (HCCPP) compounds were determined. The chemical structure of the compounds is presented in Fig. 15. The two lower homologues possess only a nematic phase, while the heptyl compound has a smectic phase in addition to a nematic phase.

The crystal and molecular data of the phenylcyclohexane pyrimidines are summarised in Table 9. The crystal structure of the ethyl compound (ECCPP) contains two crystallographically independent molecules [89]. The phenyl and pyrimidine rings of both molecules are planar. The cyclohexyl ring of molecule 1 has a chair conformation, while the cyclohexane ring of molecule 2 is approximately planar. With regard to the crystal packing of ECCPP, the molecules form a complicated structure in which the dipolar interactions of the cyano groups play a role. The contact distance between the cyano groups of adjacent molecules is around 3.40 Å. Compound PCCPP crystallises in the monoclinic space group P2/a with four molecules per unit cell [90]. The phenyl and pyrimidine rings are planar and lie almost in the *ac* plane. The cyclohexyl ring has a chair conformation. Pairs of molecules related by a centre of symmetry are packed into sheets in the *ac* plane and these sheets are stacked in an imbricated fashion along the *b*-axis. In the case of compound HCCPP [91], the phenyl and the pyrimidine rings are twisted by 15°. The cyclohexyl ring has a chair conformation. Pairs of molecules related by a centre of symmetry are packed parallel to the *a*-axis. The contact distance between the cyano groups of adjacent molecules is around 3.50 Å N_{cyano}-N'_{cyano}.

4
Liquid Crystals Containing Carboxylic or Thiocarboxylic Groups

In this section, we will neglect the crystal structures of the mesogenic perfluorinated phenyl benzoates [23–27], benzoic acids [6, 28–31], cinnamic acids [7, 32, 33], dicarboxylic acids [34, 35], and cinnamate compounds [8, 36–40]. The single crystal X-ray analyses of chiral mesogenic carboxylates are described in Sect. 6.

Table 9. Crystal and molecular data of mesogenic phenylcyclohexane pyrimidines

Compound	Ref.	R_1	Space group	Z	Dihedral angle between the phenyl and the pyrimidine rings/°	Selected intermolecular distances [Å]
ECCCP	[89]	H_5C_2-	$P2_1/n$	8	molecule 1: 15.2 molecule 2: 6.9	3.34 $N1_{cyano}-N2'_{cyano}$ 3.40 $N1_{cyano}-C2'_{cyano}$
PCCPP	[90]	$H_{11}C_5-$	$P2/a$	4	2	
HCCPP	[91]	$H_{15}C_7-$	$P2_1/c$	4	15	3.50 $N_{cyano}-N'_{cyano}$ 3.46 $N_{cyano}-C'_{cyano}$

4.1
Phenyl Benzoates

4.1.1
4′-Cyanophenyl-4-n-alkylbenzoates (CPnB), 4′-Cyanophenyl-4-n-alkoxybenzoates (CPnOB), and 4′-n-Alkoxyphenyl-4-cyanobenzoates (nOPCB)

During the last few years a large number of single crystal X-ray studies were carried out on mesogenic phenyl benzoate compounds. In the following we will present the crystal structures of the 4′-cyanophenyl-4-*n*-alkylbenzoates (CP*n*B), 4′-cyanophenyl-4-*n*-alkoxybenzoates (CP*n*OB), and 4′-*n*-alkoxyphenyl-4-cyanobenzoates (*n*OPCB). The general chemical structure of the mesogenic phenyl benzoates is presented in Fig. 16. The compounds of the three series have different phase sequences: crystal-nematic-isotropic for CP7B, CP6OB, CP7OB, CP8OB, 6OPCB and monotropicnematic for CP5B, CP4OB, CP5OB; crystal-smectic A-nematic-isotropic for CP8B, 7OPCB, and 8OPCB.

The crystal structures of the mesogenic compounds CP*n*B, CP*n*OB, and *n*OPCB were described by different authors [92–98]. The crystal and molecular data of the investigated compounds CP*n*B, CP*n*OB, and *n*OPCB are summarised in Table 10. In all the molecules, the carbonyloxy plane is approximately coplanar with the phenyl ring, to which the carbon atom of the carbonyloxy group is attached. The two phenyl rings of the molecules are twisted with exception of compound CP6OB.

The crystal structure of CP5P contains two crystallographically independent molecules which are packed in an approximately parallel imbricated mode [92]. No intermolecular contacts between the cyano groups of adjacent molecules are observed. The crystal structure of 4′-cyanophenyl-4-*n*-octylbenzoate (CP8B) shows close arrangements between the cyano groups of neighbouring molecules with a distance of 3.46 Å N_{cyano}-N'_{cyano} [95], as demonstrated in Fig. 17.

In 1995, Ibrahim et al. [96] described the crystal structure of 4′-cyanophenyl-4-*n*-butoxy-benzoate (CP4OB). The crystal structure of CP4OB is presented in Fig. 18. The molecular structure shows that the butoxy group is extended in a gauche conformation. The molecules are arranged parallel to [100]. The cyano groups of two adjacent molecules of different layers are closely arranged with a contact distance of 3.54 Å [N_{cyano}-N'_{cyano}].

For CP5OB, it was reported [97] that the molecules have exact C_S symmetry as a consequence of their location on crystallographic mirror symmetry planes. The cyanophenyl group is perpendicular to the mirror plane, whereas the other phenyl ring and the carbonyloxy group are coplanar and located in

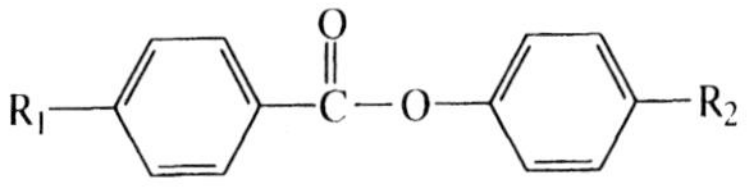

Fig. 16. Chemical structure of the mesogenic phenyl benzoates

Table 10. Crystal and molecular data of the mesogenic 4'-cyanophenyl-4-n-alkylbenzoates (CPnB), 4'-cyanophenyl-4-n-alkoxybenzoates (CPnOB), and 4'-n-alkoxyphenyl-4-cyanobenzoates (nOPCB)

Compound	Ref.	R_1	R_2	Z	Space group	Dihedral angle between the phenyl rings/°	Selected intermolecular distances [Å]
CP5B	[92]	$H_{11}C_5$-	-CN	8	$P2_1/n$	molecule 1: 60.0 molecule 2: 59.4	3.41–3.52 N_{cyano}-C'_{phenyl}[a]
CP7B	[93] [94]	$H_{15}C_7$-	-CN	4	$P2_1/n$	47.5	3.68 N_{cyano}-N'_{cyano} 3.41 N_{cyano}-C'_{cyano}
CP8B	[95]	$H_{17}C_8$-	-CN	4	$P2_1/n$	47.1	3.46 N_{cyano}-N'_{cyano} 3.48 N_{cyano}-C'_{cyano}
CP4OB	[96]	H_9C_4O-	-CN	8	Pccn	69.6	3.54 N_{cyano}-N'_{cyano}
CP5OB	[97]	$H_{11}C_5O$-	-CN	4	Pnam	90.0	3.92 N_{cyano}-N'_{cyano} 3.83 N_{cyano}-C'_{cyano}
CP6OB	[98]	$H_{13}C_6O$-	-CN	2	$P\bar{1}$	1.0	3.60 N_{cyano}-C'_{cyano}
CP7OB	[98]	$H_{15}C_7O$-	-CN	8	$P2_1/a$	molecule 1: 36.2 molecule 2: 65.3	3.43 $N2_{cyano}$-$O1'_{carbonyloxy}$ 3.29 $C2_{cyano}$-$O1'_{carbonyloxy}$ 3.38 $N1_{cyano}$-$O2'_{carbonyloxy}$
CP8OB	[98]	$H_{17}C_8O$-	-CN	2	$P\bar{1}$	49.2	3.49 N_{cyano}-C'_{cyano} 3.39 $O_{carbonyloxy}$-$O'_{carbonyloxy}$
6OPCB	[98]	NC-	-OC_6H_{13}	2	$P\bar{1}$	88.1	3.52 N_{cyano}-$C'_{carbonyloxy}$ 3.64 N_{cyano}-$O'_{carbonyloxy}$
7OPCB	[98]	NC-	-OC_7H_{15}	4	$P2_1/c$	63.1	3.52 N_{cyano}-C'_{cyano}
8OPCB	[98]	NC-	-OC_8H_{17}	4	$P2_1/c$	60.3	3.59 N_{cyano}-$C'_{carbonyloxy}$

[a] Distances to the carbon atoms of the phenyl rings.

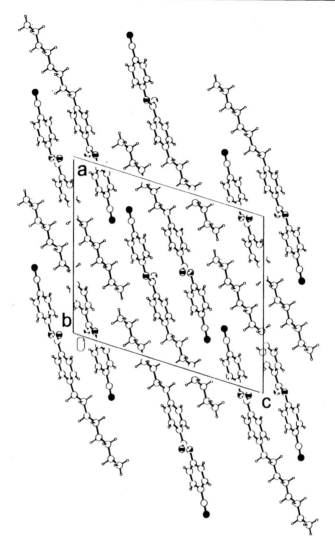

Fig. 17. Crystal structure of CP8B along [010] (Reprinted from [95])

the mirror plane. The molecules adopt a nearly stretched form. With regard to the packing of CP5OB, the angle of 35.2° between neighbouring molecules leads to a herringbone arrangement within the sheets. The crystal structure of CP5OB shows neither a parallel arrangement nor a sufficient imbrication of the molecules. This was interpreted to be responsible for the fact that the compound is only monotropic nematic. In the case of the related compound CP5B such high molecular symmetry was not found [92].

In 1995, Iki and Hori [98] presented the crystal structures of some mesogenic compounds of the two isomeric series CP*n*OB and *n*OPCB (*n* = 6,

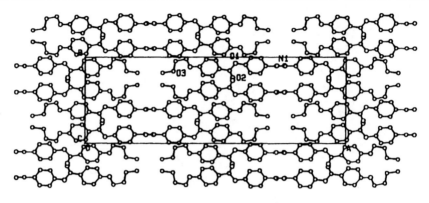

Fig. 18. Crystal structure of CP4OB along [001] (Reprinted from [96])

7, and 8). The crystal structure of CP6OB shows close contacts of the cyano groups between neighbouring molecules with a distance of 3.60 Å N_{cyano}-C'_{cyano}. The crystal structure of CP7OB contains two crystallographically independent molecules 1 and 2. Here, close contacts between the cyano and the carbonyloxy groups of the two independent molecules are observed ($N2_{cyano}$-$O1_{carbonyloxy}$ = 3.43 Å, $N1_{cyano}$-$O2_{carbonyloxy}$ = 3.38 Å). For compound CP8OB, close arrangements between the cyano groups of a pair of molecules are observed (N_{cyano}-C'_{cyano} = 3.49 Å). In addition, two carbonyloxy groups of another pair of molecules are closely arranged ($O_{carbonyloxy}$-$O'_{carbonyloxy}$ = 3.39 Å). For compound 6OPBC, the contact distances of the cyano and the carbonyloxy groups between two adjacent molecules related by an inversion centre are 3.52 Å for N_{cyano}-$C'_{carbonyloxy}$ and 3.64 Å for N_{cyano}-$O'_{carbonyloxy}$, respectively. The compounds have imbricated structures.

In contrast to the previously described compounds, the crystal structure of 7OPCB has a distinct layered structure. The cyano groups are closely arranged between antiparallel molecules, resulting in two-dimensional networks with C-N distances of 3.49–3.56 Å. The crystal structure of 8OPCB shows the formation of dimers by close contacts between the cyano and the carbonyloxy groups of adjacent molecules (N_{cyano}-$C'_{carbonyloxy}$ = 3.59 Å). The dimers have an interdigitated structure forming an alternate stacking of core moieties and chains.

Iki and Hori discussed the crystal structures of the compounds CPnOB and nOPCB in relation to the mesophase behaviour considering the dimer model proposed by Cladis et al. [99] for this kind of molecule in the liquid crystalline phase.

4.1.2
4, 4'-Disubstituted Phenyl Benzoates

In the following we will present the crystal structures of mesogenic disubstituted phenyl benzoates containing no cyano group [100–107]. The crystal and molecular data of the investigated compounds are summarised in Table 11. Only the crystal structures of some selected compounds will be discussed here.

In 1996, Hori and Nishiura [100] investigated the noncentrosymmetric crystal structure of 4-octyloxybiphenyl-4'-yl 4-methoxybenzoate. This compound shows a nematic phase. The two phenyl rings of the biphenyl group are almost coplanar with a dihedral angle of 2.6°. All the molecules are arranged in a parallel manner where the long molecular axes are oriented in one direction, resulting in a highly polar structure. The compound has an imbricated packing structure.

The crystal structure of 4-butylphenyl-4'-butylbenzoyloxybenzoate was determined by Haase et al. [101]. The compound forms a nematic phase. The neighbouring phenyl rings of the molecule are twisted by 49 and 62°. The dipole moments of the carbonyloxy groups perpendicular to the long molecular axis are compensated to each other as much as possible.

Hartung et al. [102] described the crystal structure of 4'-(β-cyanoethyl)-phenyl-4-n-pentoxybenzoate which exhibits a monotropic nematic phase. The crystal structure of the compound shows a parallel arrangement of the molecules.

Kurogoshi and Hori [104] determined the crystal structures of the mesogenic ethyl and butyl 4-[4-(4-n-octyloxybenzoyloxy)benzylidene]aminobenzoates. The compounds have different phase sequences: crystal-smectic A-nematic-isotropic and crystal-smectic C-smectic A-nematic-isotropic for the ethyl and butyl compounds, respectively. Both compounds have layer structures in the solid phase. The butyl compound contains two crystallographically independent molecules. Within the layers, adjacent molecules are arranged alternately so as to cancel their longitudinal dipole moments with each other. In the ethyl compound the core moieties are almost perpendicular to the layer plane, while in the butyl compound these moieties are tilted in the layer.

Centore et al. [106] described the crystal structures of the mesogenic bis[(4-butoxycarbonyl)-phenyl]terephthalate (B-A) and bis[(4-valeroyloxy)-phenyl]terephthalate (V-A) which contain a rigid group that can also be found in mesogenic polymers. Compound B-A shows solid state polymorphism (two solid phases I and II); the crystal structure of the solid phase I was determined. The two compounds have different phase sequences: crystal I-crystal II-smectic A-isotropic for compound B-A and crystal-nematic-isotropic for compound V-A. The crystal packing of the two compounds is different. The crystal structures of the smectogenic compound B-A and the nematogenic compound V-A are shown in Fig. 19. For compound B-A, the crystal structure of the I-phase shows the formation of molecular layers in which the long molecular axes are highly tilted with respect to the layer normal. Within each layer the molecules are strictly parallel with each other. In the solid state of the nematogenic compound V-A the molecules are packed in layers with their long molecular axes normal to the layer surface. However, the expectation of an imbricated packing of the molecules for the crystal structure of the nematogenic compound V-A and of a normal layered packing for the smectogenic compound B-A is exactly contradicted.

Table 11. Crystal and molecular data of mesogenic disubstituted phenyl benzoates

Compound	Ref.	R_1	R_2	Space group	Z
4-octyloxybiphenyl-4′-yl 4-methoxybenzoate	[100]	H_3CO-	$-C_6H_4OC_8H_{17}$	Cc	4
4-butylphenyl 4′-butylbenzoyloxybenzoate	[101]	H_9C_4-	$-COOC_6H_4C_4H_9$	P$\bar{1}$	2
4′-(-cyanoethyl)-phenyl-4-n-pentoxy-benzoate	[102]	$H_{11}C_5O-$	$-CH_2CH_2CN$	P2$_1$/n	4
4-methoxyphenyl 4′-hexyloxybenzoate	[103]	$H_{13}C_6O-$	$-OCH_3$	P2$_1$/c	4
ethyl 4-[4-(4-n-octyloxybenzoyloxy)-benzylidene]aminobenzoate	[104]	$H_{17}C_8O-$	$-CH=NC_6H_4COOC_2H_5$	P2$_1$	2
butyl 4-[4-(4-n-octyloxybenzoyloxy)-benzylidene]aminobenzoate	[104]	$H_{17}C_8O-$	$-CH=NC_6H_4COOC_4H_9$	P$\bar{1}$	4
4′-(2,2-dicyanoethyl)-phenyl 4-n-nonyloxybenzoate	[105]	$H_{19}C_9O-$	$-CH_2CH(CN)_2$	P$\bar{1}$	4
bis[(4-butoxycarbonyl)-phenyl]terephthalate	[106]	$H_3C(CH_2)_3$ $OOCH_4C_6OOC-$	$-COO(CH_2)_3CH_3$	P$\bar{1}$	1
bis[(4-valeroyloxy)-phenyl]terephthalate	[106]	$H_3C(CH_2)_3COOH_4$ C_6OOC-	$-OOC(CH_2)_3CH_3$	P2$_1$/c	2
EtOCOPhOCO(Ph)$_2$COOPhCOOEt[a]					
α-form	[107]	–	–	Pn	4
β-form		–	–	P2$_1$/c	2

a Et = ethyl; Ph = phenyl.

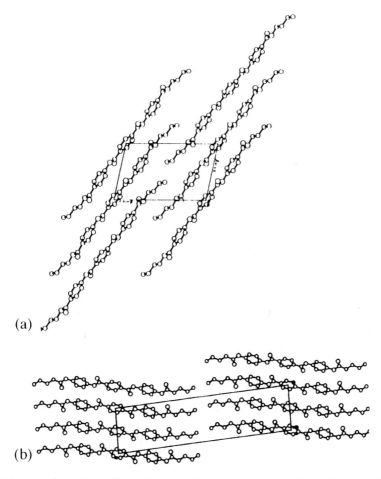

Fig. 19a,b. Crystal structure of bis[4-butoxycarbonyl)-phenyl]terephthalate (B-A) and bis[4-valeroyloxy)-phenyl]terephthalate (V-A). **a** crystal packing in B-A viewed along *a*; **b** crystal packing in V-A viewed along *b*. (Reprinted from [106])

In 1990, Tashiro et al. [107] determined the crystal structure of the mesogenic compound $H_5C_2OCO(C_6H_4)OCO(C_6H_4)_2COO(C_6H_4)COOC_2H_5$ which can be considered as model compound for liquid crystalline arylate polymers. The compound shows two crystal modifications (α-form and β-form). They found that the α-form transforms into the β-form and then into the liquid crystalline phase. The main difference in the molecular conformation of these two crystalline forms is a twist angle between the two phenyl rings of the biphenyl group: 48° for the α-form and 0° for the β-form. The crystal structure of the α-form contains two crystallographically independent molecules. The molecules are arranged in layers in which their long molecular axes are tilted by 59° from the normal to the layer surface. In the β-form the molecules are arranged in a two-dimensional layer structure, and the layers are

stacked together along the a-axis. The long molecular axes are tilted by 13°
within the layers. By Raman spectral measurements Tashiro et al. found that
the biphenyl group of the compound has a twisted structure in the liquid
crystalline state as well as in the α-form.

4.2
Phenyl Thiobenzoates

Up to now, the crystal structures of only a few mesogenic phenyl
thiobenzoates are known. The general chemical structure of the phenyl
thiobenzoates is presented in Fig. 20. The crystal structures of the two
isomeric compounds   4-n-pentylphenyl 4'-cyanothiobenzoate (NCS5) and
4-cyanophenyl 4'-pentylthiobenzoate (5SCN) were studied [108, 109]. Furthermore, the crystal structures of the mesogenic 4-hexylphenyl-4'-but-
oxythiobenzoate [110] and 4-pentyl-phenyl-4'-pentoxythiobenzoate [111]
were described. The crystal and molecular data of the investigated thiobenzoates are summarised in Table 12.

The two compounds NCS5 and 5SCN show solid state polymorphism. Both
compounds form a nematic phase. The melting points are 340 K (melting
enthalpy: 16.6 kJ/mol) for NCS5 and 350 K (melting enthalpy: 22.4 kJ/mol) for
5SCN, respectively. The investigated crystal structures of NCS5 and 5SCN are
shown in Fig. 21. In NCS5 the molecules are arranged parallel to each other
and lie perpendicular to [010] alternating in a head-to-tail configuration,
resulting in a stack formation along the b-axis. Within the layers, the
molecules are tilted [108]. The contact distances between the cyano groups of
two adjacent molecules are presented in Table 12. Moreover, a contact
distance of 3.85 Å was found between two sulphur atoms which is slightly
greater than twice the van der Waals radius of sulphur (1.85 Å). Therefore,
each molecule has two dipolar contacts.

The investigated solid phase of 5SCN crystallises in the monoclinic space
group $P2_1/c$ with 12 molecules per unit cell [109]. The three crystallographically independent molecules are in an extended form. The molecules are
arranged parallel to the (001) plane. The averaged contact distances between
the cyano groups of adjacent molecules are 3.58 Å N_{cyano}-N'_{cyano}, 3.52 Å
N_{cyano}-C'_{cyano}, and 3.89 Å C_{cyano}-C'_{cyano} (each molecule has on average five
dipolar contacts). The shortest sulphur-sulphur distance observed in 5SCN is
3.93 Å. The closer intermolecular contacts of the molecules in 5SCN compared
with NCS5 might be one reason for the higher melting point and the higher
melting enthalpy of 5SCN.

The molecular length of both compounds in the crystalline state is 19.6 Å.
X-ray investigations in the nematic phase of NCS5 showed that dimeric
(29.3 Å) as well as monomeric (18.6 Å) units exist. For SCN5, a value of 21.3 Å
was obtained for the apparent length of the molecules in the nematic phase
[108].

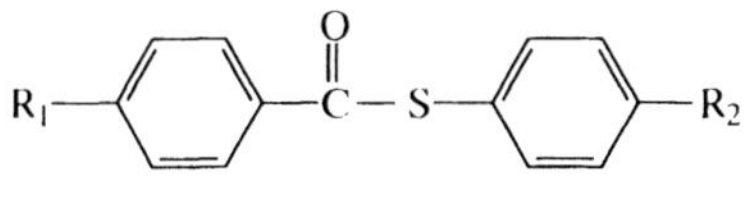

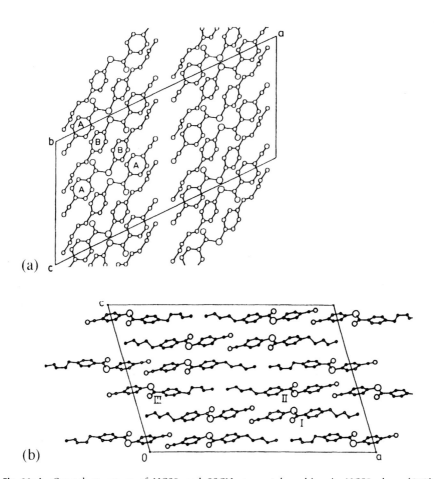

Fig. 20. Chemical structure of the mesogenic phenyl thiobenzoates

Fig. 21a,b. Crystal structure of NCS5 and 5SCN: **a** crystal packing in NCS5 along [010]; **b** crystal packing in 5SCN along [010]. (Reprinted from [108, 109])

4.3
Carboxylates

In this section, we will present a short overview of the crystal structures of different mesogenic carboxylate compounds [112–120] which are not

Table 12. Crystal and molecular data of mesogenic phenyl thiobenzoates

Compound	Ref.	R_1	R_2	Space group	Z	Dihedral angle between the phenyl rings/°	Selected intermolecular distances [Å]
NCS5	[108]	NC-	$-C_5H_{11}$	C2/c	8	69.0	3.91 $N_{cyano}-N'_{cyano}$ 3.79 $N_{cyano}-C'_{cyano}$ 3.99 $C_{cyano}-C'_{cyano}$ 3.85 S-S'
5SCN	[109]	$H_{11}C_5-$	-CN	P2$_1$/c	12	molecule I: 50.8 molecule II: 46.8 molecule III: 47.0	3.58 $N_{cyano}-N'_{cyano}$ [a] 3.52 $N_{cyano}-C'_{cyano}$ [a] 3.89 $C_{cyano}-C'_{cyano}$ [a] 3.93 S-S'
4S6	[110]	H_9C_4O-	$-C_6H_{13}$	P2$_1$/c	4	74.2	
5S5	[111]	$H_{11}C_5O-$	$-C_5H_{11}$	P2$_1$/c	4	64.8	

[a] Averaged distances.

Table 13. Crystal data and phase sequences of mesogenic carboxylate compounds

Compound	Ref.	Phase sequence	Space group	Z
4'-cyanophenyl-4-*n*-pentylcyclohexanoate	[112]	C-N-I	P$\bar{1}$	2
4-cyanophenyl-*trans*-4-[(*trans*-4-*n*-butylcyclohexyl)methyl]- cyclohexanoate	[113]	C-[N]-I	P2$_1$/n	4
2-bromo-1,4-phenylenediyl bis(*trans*-4-*n*-propyl-cyclohexanoate)	[114]	C-N-I	P$\bar{1}$	1
2,3-dicyano-4-*n*-hexyloxyphenyl-*trans*-4-*n*-heptyl-cyclohexanoate	[115]	C-[S$_A$]-I	P2$_1$/n	4
4-(2-cyanoethyl)cyclohexyl-4-*n*-pentylcyclohexane-carboxylate	[116]	C-[S$_B$]-I	P$\bar{1}$	2
4-isothiocyanato phenyl 4-pentylbicyclo[2,2,2]octane-1-carboxylate	[117]	C-N-I	P$\bar{1}$	2
trans-4-propyl cyclohexyl-4-(*trans*-4-pentylcyclohexyl)benzoate	[118]	C-N-I	P$\bar{1}$	2
4'-*n*-pentyl cyclohexyl-(4-*n*-pentyl-cyclohexane)-1-carboxylate	[119]	C-S-I	P1	1
6-hexylamino-1,2,4,5-tetrazin-3-yl 4-pentyloxybenzoate	[120]	C-N-I	P$\bar{1}$	2

C = crystalline phase: N = nematic phase; S$_A$ = smectic A phase; S$_B$ = smectic B phase; I = isotropic phase.

described in Sects. 4.1, 4.2 and 6, respectively. The crystal data and the phase sequences of the investigated compounds are summarised in Table 13.

In 1985, Baumeister et al. [113] presented the crystal structure of 4-cyano-phenyl-*trans*-4'-[(*trans*-4-*n*-butylcyclohexyl)methyl]cyclohexanoate (CBMC). The molecular structure of this compound is shown in Fig. 22. The molecules

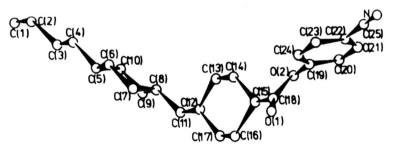

Fig. 22. Molecular structure of CBMC (Reprinted from [113])

adopt an elongated banana-shaped form. The crystal structure shows an antiparallel arrangement of the molecules into sheets. There are no intermolecular contacts between the cyano groups.

The crystal structure of 2-bromo-1,4-phenylenediyl bis(*trans*-4-*n*-propylcyclohexanoate) was determined by Hartung and Winter [114]. The molecules exhibit pseudo-centrosymmetry in consequence of a special kind of disorder within the crystal lattice. The peculiarity of the crystal structure is the disorder of the molecules with respect to the position of their bromine atoms which occupy the 2- or 5-position of the phenyl ring in a statistical manner.

Baumeister et al. [116] described the crystal structure of 4-(2-cyanoethyl)-cyclohexyl 4-*n*-pentylcyclohexanecarboxylate. The almost fully stretched molecule is only distorted by the gauche conformation of the cyanoethyl group. The crystal packing is characterised by a discrete layered arrangement with an antiparallel orientation of neighbouring molecules. With respect to intermolecular interactions, no remarkable contacts between the cyano groups can be observed.

In 1997, Sridhar et al. [117] investigated the crystal structure of the mesogenic 4-isothio-cyanato phenyl 4-pentylbicyclo[2,2,2]octane-1-carboxylate. The crystal structure of this compound shows a layered packing along the *a*-axis and imbrication along the other two axes.

Hartung et al. [120] determined the crystal structure of the mesogenic 6-hexylamino-1,2,4,5-tetrazin-3-yl 4-pentyloxybenzoate. The plane of the bridging carboxylic group is inclined to those of the phenyl and the tetrazine rings by 12.1 and 76.3°, respectively. The molecules are arranged in sheets parallel to (021).

5
Liquid Crystals Containing Azo, Azoxy, Azine, Azomethine Units or an Ethylene Bridge

5.1
Azo and Azoxy Compounds

A large number of single-crystal X-ray studies were carried out on mesogenic azo and azoxy compounds [3, 5, 121–131]. The general chemical structure of

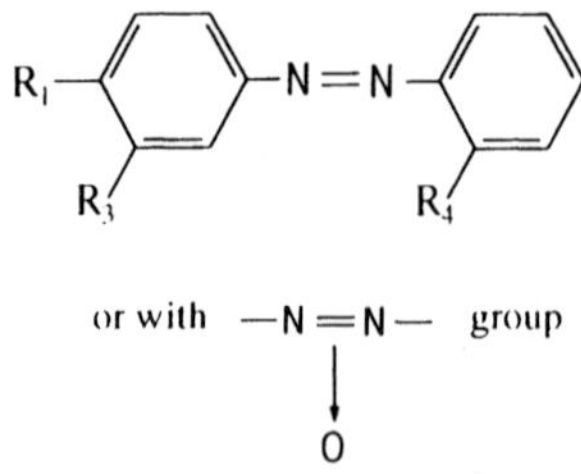

Fig. 23. Chemical structure of mesogenic azo and azoxy compounds

Table 14. Crystal data and phase sequences of mesogenic azo and azoxy compounds (see Fig. 23)

Compound	Ref.	R_1	R_2	Phase sequence	Space group	Z
N-p-methoxybenzylidene-p-phenylazoaniline	[121, 122]	H-	-N=CHC$_6$H$_4$OCH$_3$	C-N-I	P2$_1$/c	8
4,4'-azodiphenetole	[5]	H$_5$C$_2$O-	-OC$_2$H$_5$	C-N-I	P2$_1$/c	4
4-(4'-ethoxyphenylazo)phenyl valerate	[123]	H$_5$C$_2$O-	-OOCC$_4$H$_9$	C-N-I	P$\bar{1}$	2
4-(4'-ethoxyphenylazo)phenyl hexanoate	[124]	H$_5$C$_2$O-	-OOCC$_5$H$_{11}$	C-N-I	P2$_1$/c	4
4-(4'-ethoxyphenylazo)phenyl heptanoate	[125]	H$_5$C$_2$O-	-OOCC$_6$H$_{13}$	C-N-I	P$\bar{1}$	2
4-(4'-ethoxyphenylazo)phenyl undecylenate	[126]	H$_5$C$_2$O-	-OOCC$_{10}$H$_{21}$	C-N-I	P2$_1$/n	4
4-ethoxy-3'-(4-ethoxy-phenylimino-methyl)-4'-(4-methoxy-benzoyloxy)azobenzene[a]	[127]	H$_3$COH$_4$C$_6$COO-	-OC$_2$H$_5$	G$_I$-C$_{II}$-[N]-I	P$\bar{1}$	2
4-(4'-ethoxybenzoyloxy)-2-butoxy-4'-(4-butoxysalicylaldimine)azobenzene[b]	[128]	H$_9$C$_4$OH$_3$(HO)C$_6$CH=N-	-OOCC$_6$H$_4$OC$_2$H$_5$	C-N-I	C2/c	8
p-azoxyanisole (stable yellow form)	[3]	H$_3$CO-	-OCH$_3$	C-N-I	P2$_1$/a	4
4,4'-di-n-propoxyazoxybenzene	[129]	H$_7$C$_3$O-	-OC$_3$H$_7$	C-N-I	P2$_1$/n	4
4,4'-bis(pentyloxy)azoxybenzene	[130]	H$_{11}$C$_5$O-	-OC$_5$H$_{11}$	C-S-N-I	P$\bar{1}$	2
ethyl p-azoxybenzoate	[131]	H$_5$C$_2$OOC-	-COOC$_2$H$_5$	G$_I$-C$_{II}$-S-I	P$\bar{1}$	2

R_3=H and R_4=H with expectation of [a]R_3=-CH=NC$_6$H$_4$OC$_2$H$_5$; [b]R_4=OC$_4$H$_9$

the compounds is shown in Fig. 23. The crystal data and the phase sequences of the investigated mesogenic azo and azoxy compounds are presented in Table 14.

Vani and Vijayan [121] investigated the crystal structure of N-p-methoxy-benzylidene-p-phenyl-azoaniline (MBPAA). The two crystallographically independent molecules are oriented antiparallel to each other. The packing of the molecules shows a herringbone arrangement.

In 1990, Baumeister et al. [127] described the crystal and molecular structure of 4-ethoxy-3'-(4-ethoxyphenyliminomethyl)-4'-(4-methoxy-benzoyloxy)azobenzene. The molecules have a bifurcated shape. The phenyliminomethyl branch is bent markedly from the nearly linear three ring fragment, but is almost coplanar with the azobenzene moiety. They found that the molecular conformation is affected by an intramolecular interaction of the carboxylic and azomethine groups. The crystal packing was described in terms of a sheet structure with interdigitating rows of molecules.

Perez et al. [128] investigated the crystal structure of 4-(4'-ethoxybezoyloxy)-2-butoxy-4'-(4-butoxysalicylaldimine)azobenzene. This compound contains four rings in the main core and a lateral alkoxy branch on one of the inner rings. The lateral butoxy chain is nearly perpendicular to the long axis of the main core. This molecular conformation induces the molecules to make a very complex network in the solid. The crystal cohesion is due to van der Waals interactions. The change of the lateral chain conformation in the solid and nematic phases is discussed.

In 1933, Bernal and Crowfoot [1] reported on the solid state polymorphism of p-azoxyanisole. They found two crystalline modifications of this compound, a stable yellow form and an unstable white polymorph. Krigbaum et al. [3] reexamined the crystal structure of the stable yellow form. The compound shows an imbricated structure which is the basic packing required for nematic behaviour according to Gray [132].

5.2
Azines

In the last few years the crystal structures of some mesogenic azine compounds have been determined [2, 133–136]. The general chemical structure of the azines is shown in Fig. 24. The crystal data and the phase sequences of the investigated mesogenic azines are presented in Table 15.

Centore et al. [133, 135] reported on the crystal structure analyses of some mesogenic cyano-azines containing strong electron donor-acceptor groups on the phenyl rings. These mesogenic cyanoazines are simple models of

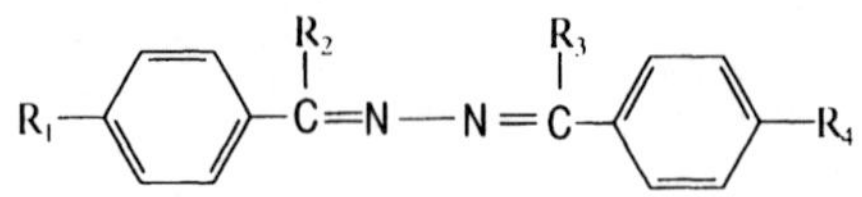

$$R_1 - \text{\textcircled{}} - \overset{R_2}{\underset{|}{C}} = N - N = \overset{R_3}{\underset{|}{C}} - \text{\textcircled{}} - R_4$$

Fig. 24. Chemical structure of mesogenic azines

Table 15. Crystal data and phase sequences of mesogenic azines (see Fig. 24)

Compound	Ref.	R_1	R_2	R_3	R_4	Phase sequence	Space group	Z
4-methoxybenzaldehyde-(4-cyanobenzylidene)- hydrazone	[133]	H_3CO-	H-	H-	-CN	C-N-I	$P2_1/n$	4
anisaldehyde azine	[2]	H_3CO-	H-	H-	$-OCH_3$	C-N-I	Cc	4
4,4'-di(N)hexoxybenzalazine	[134]	$H_{13}C_6O-$	H-	H-	$-OC_6H_{13}$	C-N-I	$P\bar{1}$	1
4-methoxyacetophenone-(4-cyanobenzylidene)-hydrazone	[133]	H_3CO-	H_3C-	H-	-CN	C-N-I	$P\bar{1}$	8
4-methoxybenzaldehyde-[(4-cyanophenyl)-ethylidene]hydrazone[a]	[133]	H_3CO-	H-	H_3C-	-CN	$C_I-C_{II}-N-I$	$P\bar{1}$	4
4-dimethylaminobenzaldehyde-[(4-cyanophenyl)-ethylidene]hydrazone	[135]	$(H_3C)_2N-$	H-	H_3C-	-CN	C-[N]-I	Pbca	8
azinobis(ethylidyne-p-phenylene) dipropionate	[136]	H_5C_2COO-	H_3C-	H_3C-	$-OOCC_2H_5$	C-N-I	$P2_1/c$	4

[a] The crystal structure of the C_{II}-phase was determined.

monomeric precursors for the synthesis of comb-like polymers with potential non-linear optical properties of the second order. All these azine compounds show a nematic phase. They found some differences in the solid state conformation of these cyanoazines due to the introduction of a methyl group onto the azine system.

In 1985, Astheimer et al. [134] described the crystal structure of the mesogenic 4,4'-di(N)-hexoxybenzalazine. Here, the centrosymmetric molecule is fully elongated with the hexoxy group in an all-trans conformation and the phenyl rings are exactly coplanar.

At the same time, Ciajolo et al. [136] presented the crystal structure of azinobis(ethylidyne-p-phenylene)dipropionate. The crystal structure contains two crystallographically independent molecules with an almost planar conformation, the phenyl rings being slightly rotated with respect to the average molecular plane. The molecules are arranged in layers.

5.3
Benzylideneanilines

During the past few years the crystal structures of many mesogenic benzylideneaniline compounds have been published [12, 137–146]. The chemical structure of the benzylideneanilines is presented in Fig. 25. The crystal data and the phase sequences of the investigated compounds are summarised in Table 16.

The compound 4-methoxybenzylidene-4'-n-butylaniline (MBBA) is a well-known thermotropic mesogenic material which is considered as a model compound for liquid crystal studies. MBBA forms a nematic phase near room temperature. On cooling, MBBA shows a large variety of solid state phases. The known solid phases of MBBA are denoted as C_0 (glassy state), I and II (relaxed amorphous states), III and IV (metastable crystalline phases), V and VI (stable crystalline phases) [147]. In 1992, Boese et al. [138] determined the crystal structure of MBBA at 110 K. MBBA crystallises in the noncentrosymmetric space group $P2_1$ with six molecules per unit cell. The three crystallographically independent molecules having slightly different conformations are arranged in layers parallel to (001). The thickness of one layer is close to the c parameter of the unit cell (18.41 Å). Boese et al. assigned this crystal structure to the V-phase of MBBA. In 1995, More et al. [139] investigated the crystal structure of the IV-phase of MBBA. By comparing the results from the powder diffraction studies with a simulated pattern given by this structure they showed that the

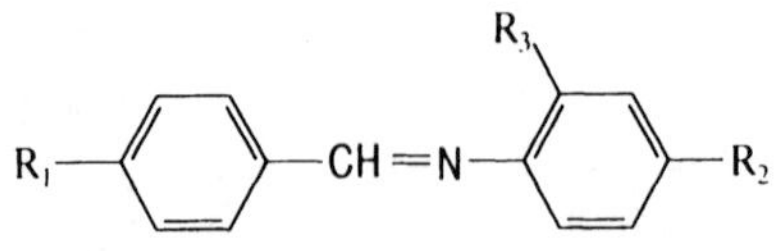

$$R_1 \underset{\text{—}}{\overset{\text{—}}{\bigcirc}} \text{—CH} = \text{N} \underset{\text{—}}{\overset{R_1}{\bigcirc}} \text{—R}_2$$

Fig. 25. Chemical structure of mesogenic benzylideneanilines

Table 16. Crystal data and phase sequences of mesogenic benzylideneanilines (see Fig. 25)

Compound	Ref.	Crystal phase	R_1	R_2	R_3	Phase sequence	Space group	Z
p-[p'-ethoxybenzylidene)amino] benzonitrile	[12]		NC-	$-OC_2H_5$	H-	C-N-I	$P2_1/c$	4
p-methoxybenzylidene-cyanoaniline	[137]		H_3CO-	-CN	H-	C-N-I	$P\bar{1}$	2
4-methoxybenzylidene-4'-n-butylaniline	[138,139]	IV	H_3CO-	$-C_4H_9$	H-	C_{I-VI}-N-I	$P2_1$	6
4-ethoxybenzylidene-4'-n-butylaniline	[137]	II	H_5C_2O-	$-C_4H_9$	H-	C_{I-IV}-N-I	$P2_1/c$	4
4-ethoxybenzylidene-4'-(4'-trifluoro-n-butyl)aniline	[140] [140]	1	H_5C_2O-	$-(CH_2)_3CF_3$	H-	C-S-I	$P\bar{1}$ $P\bar{1}$	2 6
p-[p'-methoxybenzylidene)amino] phenyl acetate	[12]		H_3CCOO-	$-OCH_3$	H-	C_I-C_{II}-N-I	$Pna2_1$	4
N-(4-n-butyloxybenzylidene)-4'-ethylaniline	[141]		H_9C_4O-	$-C_2H_5$	H-	C-S_B-I	$P\bar{1}$	2
N-(4-n-butyloxybenzylidene)-4'-octylaniline	[142]		H_9C_4O-	$-C_8H_{17}$	H-	C-S_B-I	$P\bar{1}$	4
N-(4-n-heptyloxybezylidene)-4'-hexylaniline	[141]		$H_{15}C_7O-$	$-C_6H_{13}$	H-	C-S_G-I	$P\bar{1}$	2
N-(4-n-octyloxybenzlidene)-4'-butylaniline	[143]		$H_{17}C_8O-$	$-C_4H_9$	H-	C-S_G-S_B-S_A-I	$P\bar{1}$	2
bis-(4'-n-butoxybenzal)-2-chloro-1,4-phenyline-diamine	[144]		$H_{17}C_8O-$	$-N=CHC_6H_4$ OC_8H_{17}	Cl-	C-N_C-I	$P2_1/c$	4
2-cyano-N-[4-(4-n-pentyloxy bezoyloxy)benzylidene]aniline	[145]		$H_{11}C_5$ OH_4C_6COO-	-H	NC-	C-N-I	$P\bar{1}$	2
α,ω-bis(4-n-pentylaniline benzylidene-4 oxy) butane	[146]		$H_{11}C_5H_4C_6$ $N=HCH_4C_6O$ $(H_2C)_4O-$	$-C_5H_{11}$	H-	C-S_B-S_A-I	$P\bar{1}$	2
α,ω-bis(4-n-pentylaniline benzylidene-4 oxy) pentane	[146]		$H_{11}C_5H_4C_6N$ $=HCH_4C_6O$ $(H_2C)_5O-$	$-C_5H_{11}$	H-	C-S_A-I	$P2_1$	4

crystal structure previously published by Boese et al. [138] corresponds to the IV-phase of MBBA.

Seddon and Williams [146] presented the crystal structures of α,ω-bis(4-*n*-pentylaniline-bezylidene-4'-oxy)butane (5.O4O.5) and α,ω-bis(4-*n*-pentylanilinebezylidene-4'-oxy)pentane (5.O5O.5). The molecules have even and odd chain length spacers linking the two mesogenic groups. The X-ray analysis of compound 5.O4O.5 shows that the molecules have an extended nearly linear conformation. The spacer chain O—(CH$_2$)$_4$—O of the compound has a non-planar geometry. The molecules in 5.O4O.5 are packed into strongly tilted monolayers. In contrast to 5.O4O.5, the spacer chain of compound 5.O5O.5 is essentially planar. The structure of 5.O5O.5 has two crystallographically independent molecules with an essentially linear extended chain conformation. The crystal packing of 5.O5O.5 is described in terms of a sheet structure. However, alternate pairs of sheets are displaced along the *c* direction, producing an interdigitated arrangement of layers.

5.4
Liquid Crystals Containing an Ethylene Bridge

In this section, we will report on the X-ray structure analyses of some mesogenic compounds containing an ethylene bridge. The crystal data and the phase sequences of the investigated compounds are summarised in Table 17.

In 1986, Walz and Haase [148] presented the crystal structure of the mesogenic hydrocarbon compound 1,2-bis-(4'-pentylcyclohexyl)ethane. The compound exhibits a smectic B phase over a remarkably broad range of temperature. To our knowledge, this is the only crystal structure determination of a mesogenic hydrocarbon compound up to now. Since this compound does not contain any polar groups, the arrangement in the crystalline state is

Table 17. Crystal data and phase sequences of mesogenic compounds containing an ethylene bridge and some derivatives

Compound	Ref.	Chemical structure	Phase sequences	Space group	Z
1,2-bis(4'-pentylcyclohexyl) ethane	[148]	H$_{11}$C$_5$—◯—CH$_2$—CH$_2$—◯—C$_5$H$_{11}$	C-S$_B$-I	P$\bar{1}$	1
1,2-bis(4-pentylcyclohexyl) ethane-1-ol	[149]	H$_{11}$C$_5$—◯—CH—CH$_2$—◯—C$_5$H$_{11}$ (OH)	C-S$_B$-I	P$\bar{1}$	1
1,2-bis(4-pentylcyclohexyl) ethane-1-one	[149]	H$_{11}$C$_5$—◯—C—CH$_2$—◯—C$_5$H$_{11}$ (=O)	C-S$_B$-I	P$\bar{1}$	1
1-(4'-cyanophenyl)-2-(4''-pentylcyclohexyl)-ethane	[150]	NC—◯—CH$_2$—CH$_2$—◯—C$_5$H$_{11}$	C-N-I	P$\bar{1}$	6
N,N'-bis-(p-butoxy-benzylidene)-α,α'-bi-p-toluidine	[151]	H$_9$C$_4$O—◯—〜N—◯—CH$_2$-CH$_2$—◯—N〜◯—OC$_4$H$_9$	C-S$_G$-N-I	P$\bar{1}$	2

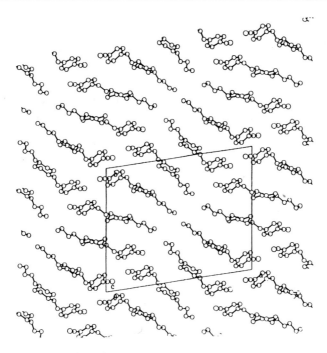

Fig. 26. Crystal structure of 1-(4'-cyanophenyl)-2-(4"-pentylcyclohexly)ethane along the *c*-axis

strongly dominated by repulsion potentials. The molecules are organised with their longest extension in parallel sheets (101).

In 1996, the crystal structures of the mesogenic 1,2-bis-(4'-pentylcyclohex-yl)ethan-1-ol and 1,2-bis-(4'-pentylcyclohexyl)ethane-1-one were described by Hoffmann et al. [149]. Both compounds show the same phase behaviour as 1,2-bis-(4'-pentylcyclohexyl)ethane. All three compounds crystallise in the tri-clinic space group P1̄ with Z = 1 and similar lattice parameters (see Table 17). For 1,2-bis-(4'-pentylcyclohexyl)ethan-1-ol, weak hydrogen bonds between the hydroxy groups have been indicated. With regard to the isostructural crystal packing of the three compounds, neither the weak hydrogen bonds in 1,2-bis-(4-pentylcyclohexyl)ethan-1-ol nor the polar carbonyl groups in 1,2-bis-(4'-pentylcyclohexyl)ethane-1-one have much influence on the crystal structure, which is also determined by van der Waals and steric interactions between the non-polar parts of the molecules.

Walz and Haase [150] investigated the crystal structure of the mesogenic 1-(4'-cyanophenyl)-2-(4"-pentylcyclohexyl)ethane. The crystal structure con-tains three crystallographically independent molecules. The crystal data of this compound are: a = 21.475 Å, b = 17.603 Å, c = 7.558 Å, α = 83.91°, β = 87.05°, γ = 80.21°, space group P1̄, and Z = 6. The crystal packing of this compound is presented in Fig. 26. A clustering of six molecules through cyano-cyano contacts in form of an island can be observed. The five averaged cyano-cyano distances of the molecules are 3.58 Å N_{cyano}-N'_{cyano}, 3.52 Å N_{cyano}-C'_{cyano}, and 3.88 Å C_{cyano}-C'_{cyano}, respectively.

In 1995, Mandal et al. [151] described the crystal structure of N,N'-bis-(p-butoxybenzylidene)-α,α'-bi-p-toluidine. They found that the two symmetrical fractions of the molecule are almost planar, but the angle between these planes is 63.5°. The molecules are arranged in tilted layers. This tilted layer-like structure is referred as a precursor to the tilted smectic phase.

6
Ferroelectric and Antiferroelectric Liquid Crystals

In this section, we will present the crystal structures of chiral mesogenic compounds exhibiting ferroelectric liquid crystalline phases which are listed in Table 18 [152–166]. Moreover, four compounds of the list show antiferro-electric properties and two compounds form only orthogonal smectic phases. The general chemical structures of the investigated chiral compounds are shown in Fig. 27.

The compounds crystallise in noncentrosymmetric space groups namely P1, $P2_1$, C2, and $P2_12_12_1$ (but with priority of $P2_1$) due to the chirality of the molecules. Most of the compounds have a tilted layer structure in the crystalline state. The tilt angle of the long molecular axes with respect to the layer normal in the crystal phase of the compounds is also presented in Table 18. Some compounds show larger tilt angles in the crystalline state than in the smectic C* phase. In the following only the crystal structures of some selected chiral liquid crystals will be discussed.

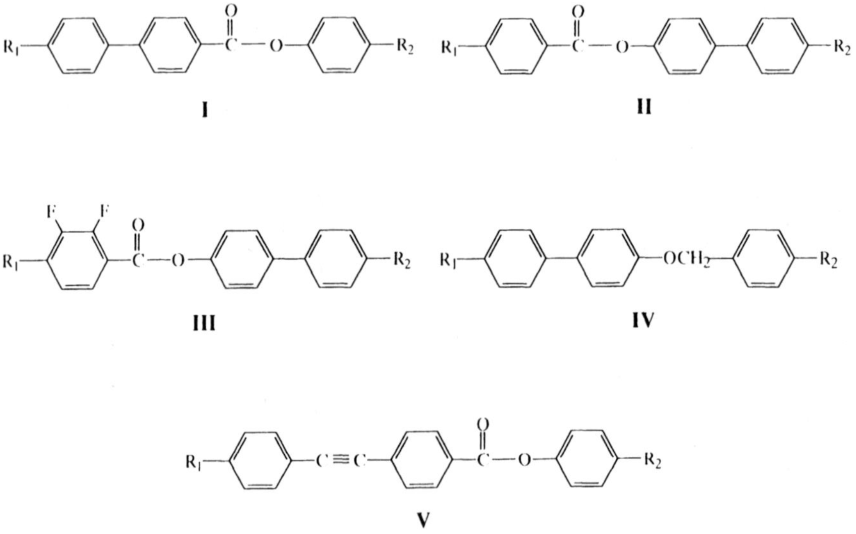

Fig. 27. Chemical structures of chiral mesogenic compounds

Table 18. Crystal data and phase sequences of chiral mesogenic compounds (see Fig. 27)

Compound	Ref.	Series	R_1	R_2	Phase sequence	Tilt angle[a] [°]	Space group	Z
4-[(S)-2-methylbutyl] phenyl 4'-hexylbiphenyl-4-carboxylate	[152]	I	$H_{13}C_6-$	$-CH_2C^*H(CH_3)C_2H_5$	$C-S_A-N^*-I$	30	$P2_1$	4
4-[(S)-2-methylbutyl]phenyl 4'-octylbiphenyl-4-carboxylate	[153]	I	$H_{17}C_8-$	$-CH_2C^*H(CH_3)C_2H_5$	$C-S_G-S_I^*-S_F^*-S_I^*-S_C^*-S_A-N^*-I$	25	$P2_1$	4
4-[(S)-2-methylbutyl]phenyl 4'-heptyloxybiphenyl-4-carboxylate	[154]	I	$H_{15}C_7O-$	$-CH_2C^*H(CH_3)C_2H_5$	$C-S_C^*-S_A-N^*-I$	30	$P2_1$	2
4-[(S)-1-methylheptyloxy] phenyl 4'-octylbiphenyl-4-carboxylate[b]	[153]	I	$H_{17}C_8-$	$-OC^*H(CH_3)C_6H_{13}$	$C-S_B-S_A-I$	15	$P2_1$	4
4-[(S)-2-methylbutyl]phenyl 4'-octyloxybiphenyl-4-carboxylate	[154]	I	$H_{17}C_8O-$	$-CH_2C^*H(CH_3)C_2H_5$	$C-S_C^*-S_A-N^*-I$	30	$P2_12_12_1$	4
4-[(S)-1-cyano-1-methylnonyl]phenyl 4'-octyloxybiphenyl-4-carboxylate[b]	[155]	I	$H_{17}C_8O-$	$-C^*(CN)(CH_3)C_8H_{17}$	$C-S_C^*-I$	30	$P1$	3
4-[(S)-1-methylheptyloxy] phenyl 4'-octyloxy-biphenyl-4-carboxylate[b]	[156]	I	$H_{17}C_8O-$	$-OC^*H(CH_3)C_6H_{13}$	$C-S_X-S_C^*-S_A-I$	30	$P2_1$	2
4-[(S)-1-methylpentyloxy carbonyl]phenyl 4'-octyloxybiphenyl-4-carboxylate[b,c]	[157] [158]	I	$H_{17}C_8O-$	$-COOC^*H(CH_3)C_4H_9$	$C-S_{CA}^*-S_A-I$	30	$P2_1$	2

Table 18. (Continued)

Compound	Ref.	Series	R₁	R₂	Phase sequence	Tilt angle[a] [°]	Space group	Z
4-[(S)-1-methylhexyloxy carbonyl]phenyl 4'-octyloxybiphenyl-4-carboxylate[b]	[157]	I	$H_{17}C_8O-$	$-COOC^*H(CH_3)C_5H_{11}$	$C-S_C^*-S_A-I$	30	$P2_1$	2
4-[(S)-1-methylheptyloxy carbonyl]phenyl 4'-octyloxybiphenyl-4-octyloxyphenyl-4-carboxylate[b,c]	[158] [159]	I	$H_{17}C_8O-$	$-COOC^*H(CH_3)C_6H_{13}$	$C-S_{CA}^*-S_C^*-S_{C_2}^*-S_A-I$	30	$P2_1$	2
4-pentyloxyphenyl 4'-[(S)-2-methylbutyl] biphenyl-4-carboxylate	[160]	I	$H_5C_2(H_3C)HC^*H_2C-$	$-OC_5H_{11}$	$C-[S_C^*]-N^*-I$	50	$P2_1$	4
4-heptyloxyphenyl 4'-[(S)-2-methylbutyl] biphenyl-4-carboxylate	[160]	I	$H_5C_2(H_3C)HC^*H_2C-$	$-OC_7H_{15}$	$C-S_C^*-N^*-I$	60	$P1$	2
p-octylphenyl 4'-[(S)-1-methylheptyloxy] biphenyl-4-carboxylate	[161]	I	$H_{13}C_6(H_3C)HC^*O-$	$-C_8H_{17}$	$C-S_C^*-S_A-N^*-I$	10	$P2_1$	2
4-octyloxyphenyl 4'-[(S)-1-methylheptyloxy] biphenyl-4-carboxylate	[156]	I	$H_{13}C_6(H_3C)HC^*O-$	$-OC_8H_{17}$	$C-S_C^*-N^*-I$	10	$P1$	4
4'-hexyloxy-4-biphenyl-4-yl p-[(S)-2-methylbutyl] benzoate I-T	[162]	II	$H_5C_2(H_3C)HC^*H_2C-$	$-OC_6H_{13}$	$C-S_I^*-S_C^*-N^*-I$	45	$P1$	4
4'-hexyloxy-4-biphenyl-4-yl p-[(S)-2-methylbutyl] benzoate II-M						60	$P2_1$	4
4'-heptyloxy-4-biphenyl p-[(S)-2-methylbutyl] benzoate II-M	[163]	II	$H_5C_2(H_3C)HC^*H_2C-$	$-OC_7H_{15}$	$C-S_I^*-S_C^*-N^*-I$	45	$P2_1$	6

4'-octylbiphenyl-4-yl p-[(S)-1-methylheptyloxy]benzoate	[161]	II	$H_{13}C_6(H_3C)HC^*O-$	$-C_8H_{17}$	$C-S_C^*-N^*-I$	65	C2	8
4-[(2S)-2-chloro-3-methylbutanoyloxy]biphenyl-4-yl 4-undecyloxy-2,3-difluorobenzoate	[164]	III	$H_{23}C_{11}O-$	$-OOCC^*HClCH(CH_3)_2$	$C-S_C^*-N^*-I$	40	$P2_1$	4
(R)-1-methylheptyl 4-(4'-octyloxybiphenyl-4-yloxymethylene)-benzoate[b,c]	[158]	IV	$H_{17}C_8O-$	$-COOC^*H(CH_3)C_6H_{13}$	$C-[S_{X2}]-S_{X1}-S_I^*-S_{CA}^*-S_C^*-I$	30	$P2_1$	2
4-(1-methylheptyloxycarbonyl)phenyl 4-heptyloxytolane-4'-carboxylate[b,c]	[165]	V	$H_{15}C_7O-$	$-COOC^*H(CH_3)C_6H_{13}$	$C-[S_{JA}^*]-[S_{IA}^*]-[S_{CA}^*]-[S_{CFI1}^*]-[S_{CFI2}^*]-S_C^*-S_{C,}-s_A-I$	23	$P2_1$	2

[a] Tilt angle of the long molecular axes with respect to the layer normal in the crystal structures.
[b] These compounds show a bent structure.
[c] The compounds show antiferroelectric properties.

Ito et al. [152] described the crystal structure of 4-[(S)-2-methylbutyl]phenyl 4'-hexylbiphenyl-4-carboxylate which shows a smectic A phase and a cholesteric phase. The molecules are arranged in a tilted smectic-like layer structure. Within the layers, the long molecular axes are tilted (30°). However, the compound exhibits no smectic C* phase.

In 1990, Colquhoun et al. [155] described the crystal structure of the ferroelectric 4-[(S)-1-cyano-1-methylnonyl]phenyl 4'-octyloxybiphenyl-4-carboxylate which shows a very high spontaneous polarisation of 520 nC/cm^2 at 50 °C. The crystal structure contains three crystallographically independent molecules with similar overall conformations. The long alkyl chain of the chiral group having all-trans conformation is almost perpendicular to the core moiety. The compound shows an interlocking layer structure in the crystal. Within a layer, the molecules are tilted. The cyano groups of the molecules are so arranged that they produce a macroscopic dipole with a substantial component parallel to the layer direction.

Since the discovery of antiferroelectric liquid crystalline phases, the interest in the synthesis of antiferroelectric materials has increased enormously in the past few years. In 1993, Hori and Endo [159] presented the crystal structure of the chiral 4-[(S)-1-methyl-heptyloxycarbonyl]phenyl 4'-octyloxybiphenyl-4-carboxylate (MHPOBC) which has been widely studied as the first example showing antiferroelectric liquid crystalline phases [167]. The investigated crystal phase of MHPOBC is metastable. The structure of the more stable crystalline phase is unknown yet. The most striking feature of this compound is that the alkyl chain of the chiral group is almost perpendicular (93°) to the core moiety. However, such a bent structure is characteristic of compounds containing a chiral 1-methylalkoxycarbonyl group. The alkyl chain of the chiral group is highly disordered with twisted conformations. The aromatic rings of the biphenyl group are approximately coplanar with the dihedral angle of 4.3°. The molecules are arranged in an interlocking smectic-like layer structure with a tilt angle close to 30° (Fig. 28). Relatively short O-O distances (3.86 Å and 3.88 Å) between the ester groups of adjacent molecules are observed. The components of the dipole moments of the polar groups are added up with respect to the 2$_1$-axis making the layers highly polar. As a precursor to the liquid crystalline state, the high polarity of the layers and the alternating arrangement of the highly polar layers were interpreted to be responsible for the antiferroelectricity of the liquid crystalline phases.

Hori and her collaborators [157, 158] also investigated the crystal structures of 4-[(S)-1-methyl-pentyloxycarbonyl]phenyl 4'-octyloxybiphenyl-4-carboxylate, 4-[(S)-1-methyl-hexyloxy-carbonyl]phenyl 4'-octyloxybiphenyl-4-carboxylate, and (R)-1-methylheptyl 4-(4'-octyloxy-biphenyl-4-yloxymethylene) benzoate. All three crystal structures are isomorphous with the structure of MHPOBC.

Hori and Ohashi [162] found three crystal forms for the compound 4'-hexyloxy-4-biphenyl p-[(S)-2-methylbutyl]benzoate. The structures of two crystal forms I-T (triclinic) and II-M (monoclinic) of this compound were determined. The crystal structure of I-T contains four crystallographically independent molecules. The dihedral angles between the phenyl rings of the

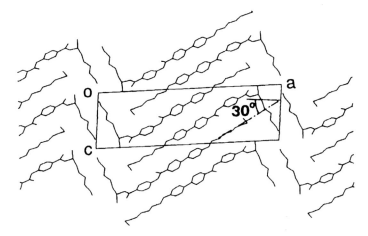

Fig. 28. Crystal structure of MHPOBC along the *b*-axis (Reprinted from [159])

biphenyl group are 21° for molecule 1, 36° for molecule 2, 42° for molecule 3, and 13° for molecule 4. All the alkyl chains have twisted conformations and those of the molecules 1, 3, and 4 are highly disordered. The chains of the chiral groups of these molecules are also disordered. The molecules are packed in a layer structure with a tilt angle of 45°. The crystal structure of II-M contains two crystallographically independent molecules. The phenyl rings of the biphenyl groups are approximately coplanar, 4.9° for molecule 1 and 4.4° for molecule 2, respectively. The molecules are arranged in layers. Within the layers, the long molecular axes are tilted (60°).

In 1996, Zareba et al. [164] presented the crystal structure of the ferroelectric 4-[(2S)-2-chloro-3-methylbutanoyloxy]biphenyl-4′-yl 4-undecy-loxy-2,3-difluorobenzoate. The two crystallographically independent molecules adopt a parallel head-to-head arrangement which gives rise to sheets parallel to the (100) plane with a thickness close to 28.5 Å. The tilt angle within the sheet is close to 40° (Fig. 29). This specific arrangement of the two independent molecules in a layer has the result that the dipole moments of the different polar groups are additive. Neighbouring sheets show alternate inclinations of the molecules generated by the 2_1 axis. The molecular arrangement is antiferroelectric-like.

Zareba et al. [165] described the crystal structure of the chiral 4-(1-methyl-heptyloxycarbonyl)-phenyl 4-heptyloxytolane-4′-carboxylate (C7-tolane) which shows monotropic antiferroelectric and ferroelectric phases. The single-crystal X-ray analysis of this compound shows that the crystal has a smectic-like layer structure composed of largely bent molecules where the chain of the chiral group is almost perpendicular (86°) to the core moiety. Within the layers, the molecules are tilted. The central tolane group of the molecule is roughly planar.

From a geometrical point of view, it is evident that all molecules having an all-trans conformation in the alkyl chain close to the chiral centre (see mark in

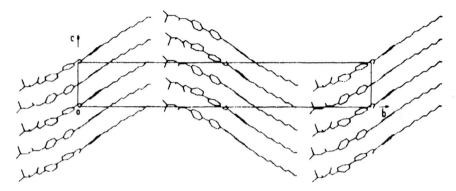

Fig. 29. Crystal structure of 4-[(2S)-chloro-3-methylbutanoyloxy]biphenyl-4'yl 4-undecy-loxy-2,3-difluorobenzoate along the x-axis (Reprinted from [164])

Table 18) must show a bent structure. This may be one reason for the interlocking and the occurrence of the 2_1-axis and the antiferroelectricity. But bent structures are possible as well as by gauche conformations like in the described solid state of compound 4-[(S)-2-methylheptyloxy]phenyl 4'-oct-ylbiphenyl-4-carboxylate [153].

7
General Discussion

A discussion of the relations between solid and liquid crystalline phases may be done in terms of energetic aspects. The phase transition enthalpy at the melting process from the crystalline state to the nematic phase (melting point) is in the range 20–30 kJ/mol, whereas the phase transition enthalpy at the clearing point (the transition from the liquid crystalline state to the isotropic phase) is smaller (<2 kJ/mol). The phase transition enthalpy of a first order character from the smectic A to the nematic phase is about 2–3 kJ/mol. At the melting process from the crystalline state to a highly ordered crystalline smectic phase the transition enthalpy is in the range 10–15 kJ/mol. The phase transition enthalpies between highly ordered smectic phases are small (2–4 kJ/mol). It is not yet clear which enthalpy or entropy contributions are needed for the realisation of the orientational, translational or positional degree of freedom.

X-ray measurements have shown that the positional and orientational ordering in highly ordered crystalline smectic phases is more comparable with the ordering in the solid state than the structures of the fluid smectic phases and the nematic phase. Therefore, a comparison of the structures of crystals and highly ordered smectic phases is possible because of their closer similarities. Up to now, only the crystal structures of few mesogenic compounds exhibiting highly ordered smectic phases have been reported in the literature [141, 143, 151, 153]. Most of the X-ray structure analyses of mesogens were carried out on compounds which form fluid smectic phases (S_A and S_C) or a nematic phase.

Concerning the question whether the detailed knowledge of the structural features in the crystalline state enables the prediction of the type and the properties of the liquid crystalline phases formed from the crystals, the following points should be mentioned.

1. It is not possible to predict from the related crystal structure alone whether the compound will melt to a liquid crystalline phase or not, because the anisotropic molecules (calamitic and discotic ones) form in favourable anisotropic packing. As a rule long shaped rod-like molecules quite often possess a layered arrangement in the solid state regardless of whether the compound is mesogenic or not.
2. An orthogonal layered structure in the solid state of rod-like molecules is the exception rather than the rule. Therefore, there is no conclusive evidence that a tilted layer structure in the solid state melts to a tilted smectic phase. In other words, if we consider the solid state as precursor for the type of the liquid crystalline state, no real precursor for an orthogonal fluid smectic phase would exist. As demonstrated in Fig. 19, the compound B-A for example exhibiting a smectic A phase has a tilted layer structure in the solid state.
3. The question of whether a certain packing in the crystalline state is a precursor for the nematic phase or for the smectic phases should be approached cautiously. As mentioned in Sect. 1, an imbricated packing of the molecules (half-to-half overlapping) in the crystalline state was considered as a precursor for the nematic phase. For instance, the nematogenic compounds CCBCH, CP7OB, CP8OB, 6OPCB possess imbricated structures in the solid state [76, 98]. However, it should be noted that in several cases the evident expectation of an imbricated packing of the molecules for the crystal structure of a nematogenic compound is exactly contradicted. As shown in Figs. 4, 18 and 19, the nematogenic compounds CBO(CH$_2$)$_7$OH, CP4OB, and V-A form layered structures in the solid state [60, 106]. On the other hand, the layered arrangement of the molecules in the solid phase was assigned to a precursor for smectic phases. In several cases smectogenic compounds possess no layer structure in the solid state, as shown in Fig. 17 for compound CP8B.
4. As the head-to-head overlapping of the molecules becomes stronger in the solid state or as more dipolar contacts between the adjacent molecules occur, the melting point of the compound is raised. In such a case the nematic phase is preferred and the observation of a smectic phase is more unlikely.
5. From the X-ray data of single crystals, it is possible to obtain information about the intermolecular interactions and the overlapping between the polar groups of neighbouring molecules. In Sects. 2.1.1, 2.1.2 and 2.1.3 the crystal structures of cyanobiphenyls were described. No cyano-phenyl overlapping of type I can be observed in the solid state of the compounds.

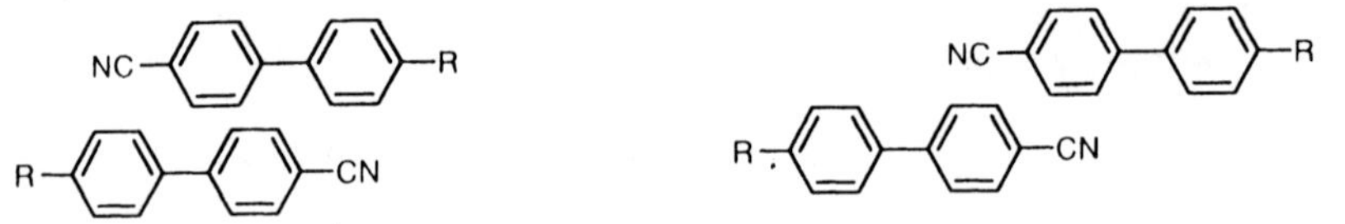

cyano-phenyl overlapping type I   cyano-phenyl overlapping type II

Fig. 30. Two types of cyano-phenyl overlapping

In some cases a cyano-phenyl overlapping of type II was found in the solid phase of the cyanobiphenyl compounds (Fig. 30).

The majority of the cyano groups attached to a phenyl ring in the para position form cyano-cyano overlapping with an averaged distance of 3.5 Å either as dimers over the centre of symmetry, islands of four, six and more groups, one dimensional chains by 2_1-axis or centre of symmetry, one dimensional bands or two-dimensional networks. X-ray measurements in the liquid crystalline state have shown that during the melting process the kind of associations change obviously. Moreover, we described the crystal structures of mesogenic compounds with more or less strong intermolecular interactions between carboxylic/thiocarboxylic groups and cyano groups as well as hydroxy groups and cyano groups. The X-ray analyses of the compounds indicate that many intermolecular contacts present in the crystals are retained in the mesophases. Of course in some cases packing energy considerations exclude dipolar contacts.

6. From X-ray measurements in the liquid crystalline phase it is impossible to determine the conformation of the molecules in the condensed state. Computer simulations give us information about the molecules' internal freedom in vacuum, but the conformations of the molecules in the condensed state can be different because of intermolecular repulsion or attraction. But it may be assumed that the molecular conformations in the solid state are among the most stable conformations of the molecules in the condensed matter and therefore also among the most probable conformations in the liquid crystalline state. Thus, as more crystallographically independent molecules in the unit cell exist, the more we can learn about the internal molecular freedom of the molecules in the condensed state.

7. Many of the investigated mesogenic compounds show solid state polymorphism. In order to obtain useful information about the arrangement of the molecules in the mesophase from the X-ray data of the single crystals, it is important to investigate the crystal structure of those solid phase which transforms into the liquid crystalline phase. For instance, only the crystal structures of the low temperature solid phases of the compounds MBBA [138, 139], MHPOBC [159], and T15 [81] could be determined, but the

structures of the solid phases close to the melting points of the compounds remain unresolved. Several research groups have tried to determine the crystal structures of the different solid phases of a compound. Interesting compounds are PCH8, PCH9, CCH5, CBO7, CBO8 and $H_5C_2OCO(C_6H_4)O$-$CO(C_6H_4)_2COO(C_6H_4)COOC_2H_5$.

Acknowledgement. We are grateful to the Deutsche Forschungsgemeinschaft for partial support of this work.

8
References

1. Bernal JD, Crowfoot D (1933) Trans Faraday Soc 29: 1032
2. Galigné JL, Falgueirettes J (1968) Acta Cryst B24: 1523
3. Krigbaum WR, Chatani Y, Barber PG (1970) Acta Cryst B26: 97
4. Krigbaum WR, Barber PG (1971) Acta Cryst B27: 1884
5. Galigné JL (1970) Acta Cryst B26: 1977
6. Bryan RF, Fallon L (1975) J C S Perkin II: 1175
7. Bryan RF, Freyberg DP (1975) J C S Perkin II: 1835
8. Cortrait M, Sy D, Ptak M (1975) Acta Cryst B31: 1869
9. Cortrait M, Destrade C, Gasparoux H (1975) Acta Cryst B31: 2704
10. Lesser DP, De Vries A, Reed JW, Brown GH (1975) Acta Cryst B31: 653
11. Bryan RF (1978) Poster Presentation at the 7th International Congress on Liquid Crystals, Bordeaux, France
12. Bryan RF, Forcier PG (1980) Mol Cryst Liq Cryst 60: 133
13. Jeffrey GA (1986) Acc Chem Res 19: 168 and references cited therein
14. Jeffrey GA, Wingert LM (1992) Liq Cryst 12: 179 and references cited therein
15. Giroud-Godquin AM, Maitlis PM (1991) Angew Chem Int Ed Engl 30: 375 and references cited therein
16. Espinet P, Esteruelas MA, Oro LA, Serrano JL, Sola E (1992) Coord Chem Rev 117: 215 and references cited therein
17. Hudson SA, Maitlis PM (1993) Chem Rev 93: 861 and references cited therein
18. Serrano JL (ed) (1996) Metallomesogens: synthesis, properties, and applications. VCH, Weinheim
19. Bideau JP, Bravic G, Cotrait M, Nguyen HAT, Destrade C (1991) Liq Cryst 10: 379
20. Allouchi H, Bideau JP, Cotrait M (1992) Acta Cryst C48: 1037
21. Nguyen HAT, Destrade C, Allouchi H, Bideau JP, Cotrait M, Guillon D, Weber P, Malthête J (1993) Liq Cryst 15: 435
22. Allouchi H, Bideau JP, Cotrait M, Destrade C, Nguyen HT (1994) Mol Cryst Liq Cryst 239: 153
23. Kromm P, Allouchi H, Bideau JP, Cotrait M (1994) Mol Cryst Liq Cryst 257: 9
24. Kromm P, Bideau JP, Cotrait M, Destrade C, Nguyen H (1994) Acta Cryst C50: 112
25. Kromm P, Allouchi H, Bideau JP, Cotrait M, Nguyen HT (1995) Acta Cryst C51: 1229
26. Kromm P, Cotrait M, Nguyen HT (1996) Liq Cryst 21: 95
27. Kromm P, Cotrait M, Rouillon JC, Barois P, Nguyen HT (1996) Liq Cryst 21: 121
28. Bryan RF, Hartley P(1980) Mol Cryst Liq Cryst 62: 259
29. Bryan RF, Hartley P, Miller RW, Shen MS (1980) Mol Cryst Liq Cryst 62: 281
30. Bryan RF, Hartley P, Miller RW (1980) Mol Cryst Liq Cryst 62: 311
31. Weissflog W, Dietzmann E, Stützer C, Drewello M, Hoffmann F, Hartung H (1996) Mol Cryst Liq Cryst 275: 75
32. Bryan RF, Forcier PG (1980) Mol Cryst Liq Cryst 60: 157

33. Bryan RF, Hartley P (1981) Mol Cryst Liq Cryst 69: 47
34. Centore R, Ciajolo MR, Roviello A, Sirigu A, Tuzi A (1990) Mol Cryst Liq Cryst 185: 99
35. Centore R, Ciajolo MR, Tuzi A (1993) Mol Cryst Liq Cryst 237: 185
36. Cotrait M, Pesquer M (1977) Acta Cryst B33: 2826
37. Leadbetter AJ, Mazid MA, Malik KMA (1980) Mol Cryst Liq Cryst 61: 39
38. Bryan RF, Hartley P (1981) Mol Cryst Liq Cryst 69: 47
39. Vani GV (1983) Mol Cryst Liq Cryst 98: 275
40. Gehring S, Fan ZX, Haase W, Müller H, Gallardo H (1989) Mol Cryst Liq Cryst 168: 125
41. Usha K, Vijayan K, Chandrasekhar S (1993) Liq Cryst 15: 575
42. Spielberg N, Luz Z, Poupko R, Praefke K, Kohne B, Pickardt J (1986) Z Naturforsch 41a: 855
43. Bonsignore S, Du Vosel A, Guglielmetti G, Dalcanale E, Ugozzoli F (1993) Liq Cryst 13: 471
44. Gray GW, Mosley A (1976) J Chem Soc, Perkin II: 97
45. Haase W, Loub J, Paulus H (1992) Z Kristallogr 202: 7
46. Haase W, Paulus H, Pendzialek R (1983) Mol Cryst Liq Cryst 100: 211
47. Vani GV (1983) Mol Cryst Liq Cryst 99: 21
48. Hanemann T, Haase W, Svoboda I, Fuess H (1995) Liq Cryst 19: 699
49. Manisekaran T, Bamezai RK, Sharma NK, Shashidhara Prasad J (1997) Liq Cryst 23: 597
50. Manisekaran T, Bamezai RK, Sharma NK, Shashidhara Prasad J (1995) Mol Cryst. Liq Cryst 268: 83
51. Manisekaran T, Bamezai RK, Sharma NK, Shashidhara Prasad J (1995) Mol Cryst. Liq Cryst 268: 45
52. Ostrovskii BI (1999) In: Mingos DPM (ed) Structure and Bonding: Liquid Crystals. Springer, Berlin Heidelberg New York
53. Walz H, Paulus H, Haase W (1987) Z Kristallgr 180: 97
54. Mandal P, Paul S (1985) Mol Cryst Liq Cryst 131: 223
55. Hori K, Koma Y, Uchida A, Ohashi Y (1993) Mol Cryst Liq Cryst 225: 15
56. Hori K, Koma Y, Kurosaki M, Itoh K, Uekusa H, Takenaka Y, Ohashi Y (1996) Bull Chem Soc Jpn 69: 891
57. Hori K, Kurosaki M, Wu H, Itoh K (1996) Acta Cryst C52: 1751
58. Yablonskii SV, Weyrauch T, Haase W, Ponti S, Strigazzi A, Veracini CA, Gandolfo C (1996) Ferroelectrics 188: 175
59. Gehring S, Quotscalla U, Paulus H, Haase W (1991) Acta Gryst C47: 2485
60. Zugenmaier P, Heiske A (1993) Liq Cryst 15: 835
61. Joachimi D, Lattermann G, Schellhorn M, Tschierske C, Zugenmaier P (1995) Liq Cryst 18: 303
62. Joachimi D, Tschierske C, Müller H, Wendorff JH, Schneider L, Kleppinger R (1993) Angew Chem Int Ed Engl 32: 1165
63. Malpezzi L, Brückner S, Galbiati E, Luckhurst GR (1991) Mol Cryst Liq Cryst 195: 179
64. Hartung H, Rapthel I (1982) Z Chem 22: 265
65. Eidenschink R, Erdmann D, Krause J, Pohl L (1977) Angew Chem 89: 103
66. Haase W, Pendzialek R (1983) Mol Cryst Liq Cryst 97: 209
67. Foitzik JK, Paulus H, Haase W (1985) Mol Cryst Liq Cryst 1: 1
68. Ibrahim IH, Paulus H, Mokhles MA, Haase W (1997) Mol Cryst Liq Cryst 299: 497
69. Paulus H, Haase W (1983) Mol Cryst Liq Cryst 92: 237
70. Haase W, Loub J, Paulus H, Mokhles MA, Ibrahim IH (1991) Mol Cryst Liq Cryst 197: 57
71. Hartung H, Rapthel I, Kurnatowski C (1981) Cryst Res Technol 16: 1289
72. Bellad SB, Babu AM, Sridhar MA, Indira A, Madhava MS, Shashidhara Prasad J (1993) Z Kristallogr 208: 43
73. Haase W, Paulus H (1983) Mol Cryst Liq Cryst 100: 111
74. Ibrahim IH, Paulus H, Haase W (1991) Mol Cryst Liq Cryst 199: 309
75. Brownsey GJ, Leadbetter AJ (1981) J Physique Lett 42: L135

76. Ji X, Richter W, Fung BM, Van Der Helm D, Schadt M (1991) Mol Cryst Liq Cryst 201: 29
77. Gupta S, Nath A, Paul S, Schenk H, Goubitz K (1994) Mol Cryst Liq Cryst 257: 1
78. Nath A, Gupta S, Mandal P, Paul S, Schenk H (1996) Liq Cryst 6: 765
79. Krieg R, Deutscher HJ, Baumeister U, Hartung H, Jaskólski M (1989) Mol Cryst Liq Cryst 166: 109
80. Walz L, Haase W, Eidenschink R (1989) Mol Cryst Liq Cryst 168: 169
81. Haase W, Paulus H, Fan ZX, Ibrahim IH, Mokhles M (1989) Mol Cryst Liq Cryst Letters 6: 113
82. Gray GW, Harrison KJ, Nash JA (1974) J Chem Soc, Chem Commun 431
83. Winter G, Hartung H, Jaskólski M (1987) Mol Cryst Liq Cryst 149: 17
84. Winter G, Hartung H, Brandt W, Jaskólski M (1987) Mol Cryst Liq Cryst 150b: 289
85. Eidenschink R, Erdmann W, Krause J, Pohl L (1980) 10th Freiburger Arbeitstagung Flüssigkristalle 2
86. Eidenschink R (1985) Mol Cryst Liq Cryst 123: 57
87. Haase W, Paulus H, Müller HT (1983) Mol Cryst Liq Cryst 97: 131
88. Walz L, Nepveu F, Haase W (1987) Mol Cryst Liq Cryst 148: 111
89. Mandal P, Paul S, Stam CH, Schenk H (1990) Mol Cryst Liq Cryst 180B: 369
90. Mandal P, Majumdar B, Paul S, Schenk H, Goubitz K (1989) Mol Cryst Liq Cryst 168: 135
91. Gupta S, Mandal P, Paul S, Wit de M, Goubitz K, Schenk H (1991) Mol Cryst Liq Cryst 195: 149
92. Baumeister U, Hartung H, Gdaniec M, Jaskólski M (1981) Mol Cryst Liq Cryst 69: 119
93. Mandal P, Paul S, Schenk H, Goubitz K (1986) Mol Cryst Liq Cryst 135: 35
94. Lokanath NK, Revannasiddaiah D, Sridhar MA, Shashidhara Prasad J, Krishne Gowda D (1997) Z Kristallogr 212: 385
95. Kilian D, Dabrowski R, Svoboda I, Fueß H, Haase W (to be published)
96. Ibrahim IH, Paulus H, Mokhles M, Haase W (1995) Mol Cryst Liq Cryst 258: 185
97. Baumeister U, Hartung H, Jaskólski M (1982) Mol Cryst Liq Cryst 88: 167
98. Iki H, Hori K (1995) Bull Chem Soc Jpn 68: 1281
99. Cladis PE, Finn PL, Goodby JW (1984) In: Griffin AC, Johnson JF (eds) Liquid crystals and ordered fluids, vol 4. Plenum Press, New York, p 203
100. Hori K, Nishiura Y (1996) Acta Cryst C52: 2922
101. Haase W, Paulus H, Pendzialek R (1983) Mol Cryst Liq Cryst 101: 291
102. Baumeister U, Brandt W, Hartung H, Jaskólski M (1983) J Prakt Chem 325: 742
103. Haase W, Paulus H, Strobl G, Hotz W (1991) Acta Cryst C47: 2005
104. Kurogoshi S, Hori K (1996) Acta Cryst C52: 660
105. Hartung H, Baumeister U, Jaskólski M (1986) Z Chem 26: 32
106. Centore R, Ciajolo MR, Roviello A, Sirigu A, Tuzi A (1991) Liq Cryst 9: 873
107. Tashiro K, Hou J, Kobayashi M, Inoue T (1990) J Am Chem Soc 112: 8273
108. Haase W, Paulus H, Ibrahim IH (1984) Mol Cryst Liq Cryst 107: 377
109. Walz L, Haase W, Ibrahim IH (1991) Mol Cryst Liq Cryst 200: 43
110. Ibrahim IH, Paulus H (1992) Mol Cryst Liq Cryst 220: 245
111. Chrusciel J, Pniewska B, Ossowska-Chrusciel MD (1995) Mol Cryst Liq Cryst 258: 325
112. Baumeister U, Hartung H, Jaskólski M (1982) Cryst Res Technol 17: 153
113. Baumeister U, Brandt W, Hartung H, Wedler W, Deutscher HJ, Frach R, Jaskólski M (1985) Mol Cryst Liq Cryst 130: 321
114. Hartung H, Winter G (1987) Cryst Res Technol 22: 259
115. Hartung H, Baumeister U, Jaskólski M, Mädicke A, Wiegeleben A (1986) Cryst Res Technol 21: 103
116. Baumeister U, Hartung H, Gdaniec M (1988) Acta Cryst C44: 1295
117. Sridhar MA, Lokanath NK, Krishnegowda D, Revanaiddaiah D, Shashidhara Prasad J (1997) Mol Cryst Liq Cryst 299: 509
118. Venkatramana Shastry CI, Babu AM, Sridhar MA, Bellad SB, Indira A, Shashidhara Prasad J (1992) Z Kristallogr 201: 177

119. Babu AM, Bellad SB, Sridhar MA, Indira A, Madhava MS, Shashidhara Prasad J (1993) Z Kristallogr 208: 49
120. Hartung H, Baumeister U, Thielemann F, Jaskólski M (1984) Acta Cryst C40: 482
121. Vani GV, Vijayan K (1977) Acta Cryst B33: 2236
122. Revannasiddaiah D, Lokanath NK, Sridhar MA, Shashidhara Prasad J (1997) Z Kristallogr 212: 387
123. Shashidhara Prasad J (1979) Acta Cryst B35: 1407
124. Shashidhara Prasad J (1979) Acta Cryst B35: 1404
125. Shaikh AM, Shivaprakash NC, Abdoh MMM, Shashidhara Prasad J (1984) Z Kristallogr 169: 109
126. Shashidhara Prasad J, Abdoh MMM, Shivaprakash NC (1983) Mol Cryst Liq Cryst 103: 261
127. Baumeister U, Kosturkiewicz Z, Hartung H, Demus D, Weissflog W (1990) Liq Cryst 7: 241
128. Perez F, Berdagué P, Judeinstein P, Bayle JP, Allouchi H, Chasseau D, Cotrait M, Lafontaine E (1995) Liq Cryst 19: 345
129. Romain F, Gruger A, Guilhem J (1986) Mol Cryst Liq Cryst 135: 111
130. Shivaprakash NC, Abdoh MMM, Shashidhara Prasad J (1985) Z Kristallogr 172: 79
131. Krigbaum WR, Barber PG (1971) Acta Cryst B27: 1884
132. Gray GW (1962) In: Molecular structure and the properties of liquid crystals. Academic Press, New York, p 186
133. Centore R, Panunzi B, Tuzi A (1996) Z Kristallogr 211: 31
134. Astheimer H, Walz L, Haase W, Loub J, Müller HJ, Gallardo M (1985) Mol Cryst Liq Cryst 131: 343
135. Centore R, Garzillo C (1997) J Chem Soc, Perkin Trans 2: 79
136. Ciajolo MR, Sirigu A,Tuzi A (1985) Acta Cryst C41: 483
137. Howard J, Leadbetter AJ, Sherwood M (1980) Mol Cryst Liq Cryst 56: 271
138. Boese R, Antipin MY, Nussbaumer M, Bläser D (1992) Liq Cryst 12: 431
139. More M, Gors C, Derollez P, Matavar J (1995) Liq Cryst 18: 337
140. Sereda SV, Antipin MY, Timofeeva TV, Struchkov YT, Shelyazhenko SV, Fialkov YA (1989) Sov Phys Crystallogr 34: 196
141. Thyen W, Heinemann F, Zugenmaier P (1994) Liq Cryst 16: 993
142. Leadbetter AJ, Mazid MA (1981) Mol Cryst Liq Cryst 65: 265
143. Gane PAC, Leadbetter AJ (1981) Mol Cryst Liq Cryst 78: 183
144. Mandal P, Paul S, Schenk H, Goubitz K (1992) Mol Cryst Liq Cryst 210: 21
145. Baumeister U, Hartung H, Gdaniec M (1987) Acta Cryst C43: 1117
146. Seddon JM, Williams DJ (1995) Liq Cryst 18: 761
147. Rosta L, Kroó N, Dolganov VK, Pacher P, Simkin VG, Török GY, Pépy G (1987) Mol Cryst Liq Cryst 144: 297
148. Walz L, Haase H (1986) Mol Cryst Liq Cryst Letters 4: 53
149. Hoffmann F, Baumeister U, Hartung H (1996) J Mol Struct 374: 373
150. Walz L, Haase H (unpublished results)
151. Mandal P, Paul S, Goubitz K, Schenk H (1995) Mol Cryst Liq Cryst 258: 209
152. Ito K, Endo K, Hori K, Nemoto T, Uekusa H, Ohashi Y (1994) Liq Cryst 17: 747
153. Ito K, Hori K (1995) Bull Chem Soc Jpn 68: 3347
154. Hori K, Takamatsu M, Ohashi Y (1989) Bull Chem Soc Jpn 62: 1751
155. Colquhoun HM, Dudman CC, O'Mahoney CA, Robinson GC, Williams DJ (1990) Adv Mater 2: 139
156. Kurogoshi S, Hori K (1997) Liq Cryst 23: 127
157. Hori K, Kawahara S, Ito K (1993) Ferroelectrics 147: 91
158. Hori K, Kawahara S (1996) Liq Cryst 20: 311
159. Hori K, Endo K (1993) Bull Chem Soc Jpn 66: 46
160. Hori K, Ohashi Y (1988) Bull Chem Soc Jpn 61: 3859
161. Hori K, Ohashi Y (1991) J Mater Chem 1: 667
162. Hori K, Ohashi Y (1991) Liq Cryst 9: 383

163. Hori K, Ohashi Y (1989) Bull Chem Soc Jpn 62: 3216
164. Zareba I, Allouchi H, Cotrait M, Nabor, MF, Destrade C, Nguyen HT (1996) Liq Cryst 21: 565
165. Zareba I, Allouchi H, Cotrait M, Destrade C, Nguyen HT (1996) Acta Cryst C52: 441
166. Hori K, Ohashi Y (1991) Mol Cryst Liq Cryst 203: 171
167. Chandani ADL, Ouchi Y, Takezoe H, Fukuda A, Terashima K, Furukawa K, Kishi A (1989) Jpn J Appl Phys 28: L1261

Packing and Molecular Conformation, and Their Relationship with LC Phase Behaviour

Boris I. Ostrovskii

Institute of Crystallography Academy of Sciences of Russia, Leninsky pr. 59, Moscow, 117333 Russia
E-mail: lev@glas.apc.org

Current studies of thermotropic liquid crystals reveal a remarkable variety of smectic phases, distinguished by their molecular packing, symmetry of one and two-dimensional lattices and tilting. We review here the structure and phase behaviour of particular types of layering with emphasis on their relationship to asymmetry of the molecular structure and conformational mobility of different molecular moieties. The breaking of up-down symmetry in the orientation of the heads (tails) of molecules is considered as having either a steric or polar origin, or as induced by the polyphilic nature of molecules. The structure of tilted smectic phases and phases with alternating layer to layer tilt are discussed in detail. We consider also the possibility of transitions from the phases with one-dimensional periodicity (smectics) to columnar phases with two-dimensional positional order. The relevance of various types of molecular model that predict the regions of stability of different layered phases are discussed.

Keywords: Smectic liquid crystals, Molecular conformations, polar and steric frustrations, Polyphilic and perfluorinated mesogens, X-ray diffraction

Structure and Bonding, Vol. 94
© Springer Verlag Berlin Heidelberg 1999

1
Introduction

Liquid crystals exhibit a rich variety of phases with orientational and partial translational order, including tilted structures. A fundamental and still unanswered question in the field of liquid crystals (LC) may be formulated in the following way: which features of molecular structure and interactions are responsible for the stability of the different LC phases? This question remains a difficult and controversial one. Apart from the evident basic characteristics, such as the strongly anisotropic shape of the molecules (of the type of rods or discs) and their polarizability anisotropy, there are a number of factors exerting a strong influence on the packing and the phase behaviour of LCs. These include the flexibility of different molecular moieties, the location and orientation of permanent dipoles and multipoles, the asymmetry of molecular shape, the polyphilic nature of mesogenic molecules, biaxiality and the effects of substitution of atoms and molecular groups in terminal or lateral positions. In this article we will not be able to address all of these factors. However, we make an attempt to overview the general tendencies in the formation of diverse layered and modulated structures formed by polar, sterically asymmetric and polyphilic molecules with different degrees of conformational freedom. We shall not give the details of experiments or theoretical considerations, but rather try to give a critical evaluation of the state of the art in the field. Emphasis is on the properties of smectic A and tilted phases with fluid layers (neither hexatic nor crystalline). On the experimental side, the results of recent X-ray diffraction and IR dichroism studies are discussed.

We start with some elementary information about anisotropic intermolecular interactions in liquid crystals and molecular factors that influence the smectic behaviour. The various types of molecular models and commonly accepted concepts reproducing the smectic behaviour are evaluated. Then we discuss in more detail the breaking of head-to-tail inversion symmetry in smectic layers formed by polar and (or) sterically asymmetric molecules and formation of particular phases with one and two dimensional periodicity. We then proceed with the description of the structure and phase behaviour of terminally fluorinated and polyphilic mesogens and specific polar properties of the achiral chevron structures. Finally, different possibilities for bridging the gap between smectic and columnar phases are considered.

2
Liquid Crystalline Phases of Thermotropics

2.1
Role of Competing Interactions and Molecular Conformations

In this section we will summarize the known results of molecular theories and computer simulations of rigid, symmetric mesogenic molecules, which will serve as a reference point for latter discussion [1]. Liquid crystals are composed of relatively large molecules that possess a strongly anisometric shape. There exist a number of anisotropic interactions between such molecules that can be responsible for the orientational order and the smectic layering. A priori, two contributions are expected to stabilize the mesophase: attractive interactions and anisotropic hard-core repulsions. In thermotropic mesogens the transitions are temperature driven and therefore some attractive interaction must be involved. The corresponding molecular theory of nematic ordering, based on the anisotropic dispersion (or Van der Waals) interactions was originally proposed by Maier and Saupe [2]. The effective potential that favours orientational order appears to be rather simple:

$$V_{eff}(\beta) \approx -Er_{ij}^{-6}(\Delta\alpha)^2 P_2(\cos\beta) \tag{1}$$

where $\Delta\alpha$ is the anisotropy of the molecular polarizability, $P_2(\cos\beta)$ is the second Legendre polynomial, β is a polar angle, characterizing the deviation of the long molecular axes from the director and E is the average excitation energy. The interaction energy (Eq. 1) depends on the magnitude of the anisotropy in the molecular polarizability $\Delta\alpha$ and is expected to be very weak for molecules with low dielectric anisotropy. Such molecules, therefore, are not supposed to exibit a nematic phase. However, there exist compounds (for example, cyclohexylcyclo-hexanes [3]) which form a stable nematic phase, but possess very low anisotropy in the molecular polarizability [4]. Moreover, the estimate of the transition temperature to the isotropic phase yields a value that is an order of magnitude too small. This clearly indicates that anisotropic dispersion forces do not make a major contribution to the stability of the orientational order. On the other hand, in accordance with the results of computer simulations [5], the nematic ordering in an athermal system of rods is formed when the effective length-to-width ratio L/D is more than three. Thus, any dense fluid composed of such molecules must be in the nematic phase at all temperatures. In reality, however, some additional contribution from attractive forces is required to stabilize the nematic phase [6].

As has been noticed by Gelbart and Gelbart [7], the predominant orientational interaction in nematics results from the isotropic dispersion attraction modulated by the anisotropic molecular hard-core. The anisotropy of this effective potential comes from that of the asymmetric molecular shape. The coupling between the isotropic attraction and the anisotropic hard-core repulsion is represented by the effective potential

$$V_{eff}(1,2) = J_{att}(r_{12})\Theta(r_{12} - \xi_{12}) \tag{2}$$

where the step-function $\Theta(r_{12} - \xi_{12})$ determines the steric cut-off. The potential (Eq. 2) can be averaged over all orientations of the intermolecular vector and can then be expanded in Legendre polynomials. The first term of the expansion has the same structure as the Maier-Saupe potential (Eq. 1), but with the coupling constant J determined by the anisotropy in the molecular shape and the average molecular polarizability α, rather than by the polarizability anisotropy. The introduction of a steric cut-off in Eq. (2) is equivalent to taking into account the short range steric correlations between rigid molecules. These correlations contribute not only to the internal energy, but also to the entropy of the system. Excluded volume effects restrict the molecular degrees of freedom and therefore the total entropy of the system is reduced. This additional contribution to the free energy is called the packing entropy and plays an important role in stabilizing the conformationally flexible and asymmetric mesogens.

Smectic A and C phases are characterized by a translational order in one dimension and a liquid-like positional order in two others. In the smectic A phase the molecules are oriented on average in the direction perpendicular to the layers, whereas in the smectic C phase the director is tilted with respect to the layer normal. A simple model of the smectic A phase has been proposed by McMillan [8] and Kobayashi [9] by extending the Maier-Saupe approach for the case of one-dimensional density modulation. The corresponding mean field, single particle potential can be expanded in a Fourier series retaining only the leading term:

$$V(z) \approx -V[1 + \alpha \cos(q_0 z)] \qquad (3)$$

where z is the direction along the layer normal, q_0 is the wave number of the layer structure ($q_0 = 2\pi/d$, d is the layer spacing) and $\alpha = 2\exp[-(\pi r_0/d)^2]$. The parameter α characterizes the strength of interaction that induces the smectic ordering and depends on the ratio between the length of the central cores of molecules r_0, where attractive interactions are localized, and that of whole molecule, including the tail chains $d \approx L$. For small values of r_0/d, the attractive interaction is strong enough and smectic phase undergoes a transition directly to the isotropic liquid. For larger values a nematic phase appears which is separated from the smectic by a second order phase transition line.

The behaviour predicted by the McMillan model is qualitatively in agreement with observations for homologous series of relatively simple, symmetrical molecules [1, 10]. These predictions, however, markedly contradict the phase diagrams of LCs consisting of asymmetric molecules [1a, 11–13]. The model potential (Eq. 3) is of the attraction type and, therefore, the McMillan theory does not outline the role of hard-core repulsions in the stabilization of the smectic phase. This appears to be particularly important after it has been shown in computer simulations by Stroobants et al. [14] that a smectic A phase can be formed in the system of hard spherocylinders at sufficiently high packing fractions. This gives clear evidence that for a smectic phase to appear there is no specific attractive interaction required – it can be formed as a result of just the steric repulsions of strongly anisometric molecules. At first sight this result seems surprizing. However, it was confirmed by subsequent density-functional theories [15]. For thermotropic systems, of course, the smectic A phase appears with decreasing temperature rather than increasing density and the inclusion of the long range intermolecular attraction is certainly required [16]. Computer simulations performed with the Gay-Berne potential also indicate the importance of the attractive tail of the potential in stabilizing of the smectic phase [17].

Thus far we have considered the rigid mesogenic molecules. The inclusion of molecular flexibility is of crucial importance for understanding the stability of various LC phases. The vast majority of liquid crystalline compounds are partially flexible due to the rotational isomerization of one or more alkyl chains. For thermotropic LCs of low molar mass the chains are usually at the terminal positions of rigid aromatic core, although the relatively long alkyl chains may also be attached to both terminal and lateral positions to give so-called phasmidic [18] and biforked mesogens [19, 20]. Chains can also be used to link two mesogenic units together and, as with main chain LC polymers, the properties of these materials strongly depend on the length of the flexible spacer and its parity [21]. In addition, chains are used in side chain LC polymers to attach mesogenic units to the polymer backbone [22].

The alkyl chains are flexible since it costs a finite, but easily achievable, energy (of the order of kT) to make rotations about any carbon—carbon bond in a given tail chain. This means that there is an appreciable fraction of *gauche* rotational states in *n*-alkyl tail chains. The populations of the various conformers cannot be measured directly, although this information can be extracted from the orientational order parameters determined at different positions along the alkyl chains, by using deuterium NMR [23]. The role of terminal chains in stabilizing the mesophase is not so obvious. On the first sight, rigid elongated molecules formed, for example, by several phenyl rings attached to each other, may be regarded as perfect candidates for LCs. However, these mesogens have very high clearing points and short temperature range for the mesophase. For example, for quinquephenyl, T_{NI} is about 430 °C and it crystallizes at T $\approx$ 380 °C [1a, 4]. The role of the flexible chains is therefore two-fold. Firstly, they provide enough conformational entropy (disorder) to prevent the total crystallization of the cores, thus expanding the range of stability of the mesophase. Secondly, they have an effect of lowering the crystal-to-mesophase and mesophase-to-isotropic liquid transition temperatures. At the crystallization point the terminal chains are frozen into the all-*trans* rotational state. Thus tail chain flexibility is essential to differentiate real smectic phases from crystalline solid phases [24]. The effects of molecular tail flexibility are clearly demonstrated in dramatic odd-even effect in nematic-isotropic and crystallization transition temperatures and transition entropies, as well as in segmental orientational order parameter variations [25] (Fig. 1).

On the theoretical side, Marcelja [26] was first to account explicitly for flexible tail chains in nematic ordering, using the Maier-Saupe model potential (Eq. 1) for each segment of the molecule. More complex models were proposed by Samulski et al. [27] and Emsley et al. [28]. In these approaches alkyl chains are assumed to exist in a discrete set of conformers described by

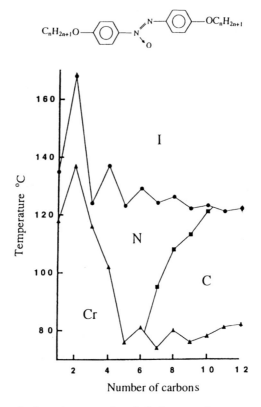

Fig. 1. Phase diagram for homologous series of alkyloxyazoxybenzenes

the rotational isomeric state model of Flory. The model [28] takes into account the statistical weights for different conformers and the energy difference between the *gauche* and *trans* states. The theory successfully describes the variation of the order parameters for the methylene groups along an alkyl chain and the odd-even effect in nematics.

A lattice model of uniaxial smectics, formed by molecules with flexible tails, was recently suggested by Dowell [29]. It was shown that differences in the steric (hard-repulsive) packing of rigid cores and flexible tails – as a function of tail chain flexibility – can stabilize different types of smectic A phases. These results explain the fact that virtually all molecules that form smectic phases (with only a few exceptions [1a, 4]) have one or more flexible tail chains. Furthermore, as the chain tails are shortened, the smectic phase disappears, replaced by the nematic phase (Fig. 1).

2.2
Polar and Sterically Asymmetric Molecules

The aforesaid considerations are valid for relatively simple symmetrical mesogenic molecules. However, there are many curious examples of asymmetric LCs (with one terminal chain or tail chains of different length and nature) where the smectic phase disappears as a result of simple reversing of the position of tail chain relative to the asymmetric central core [1a, 4].

Moreover, the new smectic phases with the broken head-to-tail inversion symmetry (of the type of A_d, A_2, $\bar{A}$, etc.) and the re-entrant effects (i.e. the "re-entrance" of the less ordered phase at temperatures below those of a more ordered phase) were found among strongly asymmetric mesogens [11–13, 30] (Fig. 2). These observations indicate that with asymmetric molecules, dipolar and steric interactions are important for the liquid crystalline behaviour.

The polar asymmetry of mesogens is usually associated with the presence of strong dipoles (typically a cyano or nitro group) at one of the ends of the molecule (Fig. 2b). However it may also be due to the difference in the number and type of cyclic and bridge fragments and the terminal groups in the molecule. For example, polar asymmetry may be introduced by replacing one of the benzene rings in a mesogenic molecule by a cyclohexane ring [31]. The interaction energy of the permanent dipoles of molecules may be fairly large: $U_{\text{dip}} \approx p^2/a^3 \approx kT/2$ for two neighbouring molecules, at distances of around 5 Å, having a longitudinal dipole moment $p \approx 5D$. In spite of a considerable value for the dipole–dipole interaction in mesogens possessing large permanent dipoles, it does not give a direct contribution to the molecular field of a system with a non polar symmetry [1b]. However, it does give rise to strong short-range dipolar correlations between molecules, including the formation of dimers, trimers etc., which play an important role in stabilizing the nematic and smectic phases composed of polar molecules [32].

For mesogenic molecules, it is not only the magnitude and location of the dipole moment but also the spatial distribution of charges due to conjugation between molecular fragments which are important for LC behaviour. Most bridge fragments of mesogenic molecules contain double and triple bonds and, consequently, they are able to realize π-electron conjugation with one another or with the nonsaturated cyclic moieties of the core, for example, phenyl rings

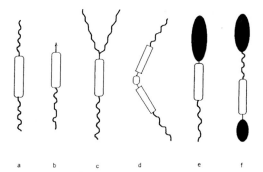

a   b   c   d   e   f

Fig. 2a–f. Mesogenic molecules with differing degrees of polar and sterical asymmetry: **a** symmetric molecule with rigid core and two hydrocarbon tails; **b** terminally polar molecule (*arrow* indicates the permanent dipole); **c** swallow-tailed (biforked) molecule; **d** banana shaped molecule; **e** terminally fluorinated molecule; **f** polyphilic molecule (*hatched areas* correspond to the fluorinated fragment)

[33]. The properties of polarized molecules depend strongly not only on the terminal dipoles but also on the magnitude and direction of the secondary molecular dipoles associated with the bridge groups and the oxygen of alkoxy chains. Reversal of the bridge groups affects the magnitude and length of conjugated molecular dipoles and can change the character of the packing of the molecules in smectic layers.

Many of the mesogenic molecules are sterically asymmetric, which is determined by the fractures and bending of the molecular core as well as by the presence of the tail chains of different nature, including the branched, biforked or polyphilic moieties (Fig. 2c–f). In terms of the multipole model of molecular asymmetry introduced by Petrov and Derzhanski [34], we can speak about longitudinal or transverse steric dipoles or multipoles (Fig. 3).

The influence of the steric dipoles on the stability of LC phases has not been studied so well. However, there are a number of strong effects in liquid crystals which are mainly determined by the strength of steric dipoles. These are the dipolar flexoelectric effect, i.e. the appearance of macroscopic polarization due to the asymmetric packing of wedge- or banana-shaped molecules induced by the orientation deformations [35], and ferroelectric ordering in chiral smectic C liquid crystals [36]. It should be noted that the division between the steric and dipolar asymmetry is not all that strong. Any steric difference between the head and tail of a molecule also necessarily implies its polarity. Nevertheless, this approach is useful to determine the origin of particular types of layering.

The combination of polar and steric dipoles and multipoles, together with the conformational flexibility of the tail chains, leads to the large variety of smectic phases [11–13]. In addition to the monolayer smectic A_1 phase, where the molecules are oriented randomly up and down within each layer, there are

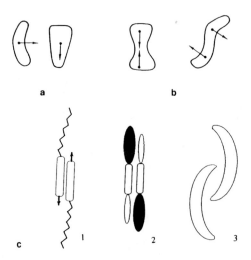

Fig. 3. a Schematic representation of mesogenic molecules possessing steric dipoles. **b** Schematic representation of mesogenic molecules possessing steric quadrupoles. **c** The formation of spatial associates of dipolar (1) and steric (2, 3) origin

bilayer smectics A_2 having a period of d $\approx$ 2L and partially bilayer (over-lapped) smectics A_d, whose layer periodicity d is incommensurate with the length L of the molecules: L < d < 2L (Figs. 3 and 4). The d in designation of this phase stands for dimers, thus underlying the importance of short-range antiparallel packing in its stability. There are also bilayer smectic phases with an alternation of the positions of the polar heads of the molecules – the antiphase or $\tilde{A}$ and the ribbon phase or $\tilde{C}$. They differ in the symmetry of their two-dimensional lattices which were found to be centered rectangular ($\tilde{A}$) and oblique ($\tilde{C}$).

These structures were firstly observed for terminally polar mesogens [11, 12]. However, recent experiments give clear evidence of the presence of smectic A_d layering [37, 38], re-entrant nematic behaviour [39], two-dimensional lattices [40, 41] and smectic layering with incommensurate periodicities [42] for non-polar sterically asymmetric LCs.

2.3
Experimental Methods

In this section we will discuss in some detail the application of X-ray diffraction and IR dichroism for the structure determination and identification of diverse LC phases. The general feature, revealed by X-ray diffraction (XRD), of all smectic phases is the set of sharp (00n) Bragg peaks due to the periodicity of the layers [43]. The in-plane order is determined from the half-width of the in-plane (hk0) peaks and varies from 2 to 3 intermolecular distances in smectics A and C to 6–30 intermolecular distances in the hexatic phase, which is characterized by six-fold symmetry in location of the in-plane diffuse maxima. The lamellar crystalline phases (smectics B, E, G, I) possess sharp in-plane diffraction peaks, indicating long-range periodicity within the layers.

For the purpose of phase identification, the well aligned samples of LCs are required. Such an alignment is usually achieved using either the magnetic field or orienting action of the free surface of the thin LC films. Moreover, many of the mesogenic compounds, on cooling from isotropic phase, form spontane-ously the well aligned homeotropic films on the clean polished substrates [44, 45]. For certain experiments, for instance to probe the in-plane short-range order of smectics or for a routine measurements of the layer periodicity, low resolution set-ups with the $\Delta q \approx 10^{-2}$ Å^{-1} (full width at half maximum, FWHM) are relevant. The diffraction vector q is defined as $q = (4\pi/\lambda) \sin\theta$ (θ and λ are the scattering angle and the wavelength of the X-rays, respectively). For other cases, which require a more precise measurement of the layer spacing or determination of correlation lengths of the order of 300–1000 Å, experimental schemes with a medium resolution $\Delta q \approx 10^{-3}$ Å^{-1} are used. These schemes utilize the focussing graphite or Ge monochromators and a linear position sensitive detector to record the diffraction maxima. The correlation length ξ or the characteristic size of the domains in which molecules scatter radiation coherently is determined from the width of the diffraction peak: $\xi = 2/\text{FWHM}$ for a Lorentzian lineshape, for example. To determine the larger correlation lengths or to analyse accurately the profiles of the XRD diffraction peaks, high resolution set-ups (with $\Delta q \approx 10^{-4}$ Å^{-1}) are needed. Such experiments require the high flux X-ray sources and the pair of perfect crystals (which can be Si (111) or Si (220) or Ge of the same orientation) arranged in the so-called dispersion free geometry [43].

The XRD patterns for the perfectly aligned smectic phases in the direction along the layer normal are shown in Fig. 4. In monolayer A_1 phase reciprocal space consists of the (00n) Bragg spots with spacing equal to $q_0 = 2\pi/d$. A bilayer A_2 phase exhibits two Bragg peaks at q_0 and $q_0/2$. The partially bilayer A_d phase is characterized by a Bragg peak at the position intermediate between q_0 and $q_0/2$. The two-dimensional phases of the type of $\tilde{A}$ and $\tilde{C}$ are identified by the appearance of the off-axis (0kn) reflections, indicative of the additional density modulation in the plane of layers [11–13]. We note that from the symmetry viewpoint the $\tilde{A}$ and $\tilde{C}$ phases belong more to columnar phases than to smectics – they are both periodic in two dimensions. However, in contrast to conventional columnar phases of disc-like molecules, which show a large number of (0kn) reflections, the $\tilde{A}$ and $\tilde{C}$ phases exhibit the limited number of combined (0kn) reflections and the absence of equatorial (0k0) diffraction spots [46, 47]. This means that the non-uniformity of the electron density is substantially less within a layer than across it. This allows one to consider the $\tilde{A}$ and $\tilde{C}$ phases formed by terminally polar mesogens together with the conventional smectic phases with one-dimensional periodicity. However, the

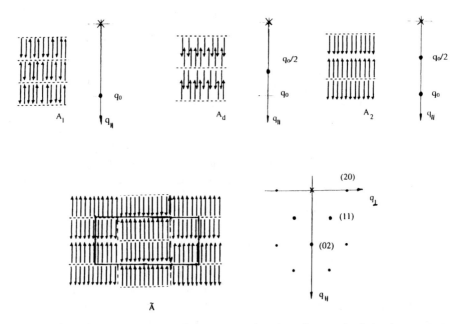

Fig. 4. Schematic representation of the smectic layering along with their characteristic diffraction patterns for the monolayer (A_1), the partially bilayer (A_d), the bilayer (A_2) and the two-dimensional ($\tilde{A}$) phases. The *arrows* indicate permanent dipoles, the *solid points* are Bragg reflections

situation might be different for other rod-like mesogens exhibiting two-dimensional lattices. For example, clear evidence of the columnar ordering has been found for the perfluorinated swallow-tailed compounds [41] (see below). This was indicated by the appearance of the sharp equatorial (0k0) and a large number of combined (0kn) reflections.

The important information about the properties of smectic layers can be obtained from the relative intensities of the (00n) Bragg peaks. The electron density profile along the layer normal is described by a spatial distribution function $\rho(z)$. The function $\rho(z)$ may be represented as a convolution of the molecular form factor $F(z)$ and the molecular centre of mass distribution $f(z)$ across the layers [43]. The function $F(z)$ may be calculated on the basis of a certain model for layer organization [37, 48]. The distribution function $f(z)$ is usually expanded into a Fourier series: $f(z) = \sum \tau_n \cos(nq_0 z)$, where the coefficients $\tau_n = \langle \cos(nq_0 z) \rangle$ are the de Gennes-McMillan translational order parameters of the smectic A phase. According to the convolution theorem, the intensities of the (00n) reflections from the smectic layers are simply proportional to the square of the translational order parameters τ_n

$$I_n \sim F^2(nq_0)\, \tau_n^2 \tag{4}$$

where $F(nq_0)$ is the Fourier transform of the molecular form factor $F(z)$. The ratio of the succesive intensities (for example, I_{002}/I_{001}) is given by

$$I_{002}/I_{001} = 1/2F^2(2q_0)\, \tau_2^2/F^2(q_0)\tau_1^2. \tag{5}$$

Once the wavelength dependence of the molecular form factor $F(nq_0)$ is known from the reasonable model of layer organization, the ratios τ_n/τ_1 may be calculated. The value of these ratios (for example, τ_2/τ_1, τ_3/τ_1) give a good guide to the sharpness of the distribution function $f(z)$ – for an ideal crystal $f(z)$ would be an array of delta-functions and $\tau_2 = \tau_1 = \cdots = \tau_n = 1$. From the expression at Eq. (5), only the relative values τ_n/τ_1 may be determined. Because of the large uncertainty of the calibration of the XRD intensity to an absolute scale, it is impossible to determine the exact values of τ_n. However, they may be estimated by assuming some analytical form for the function $f(z)$. The simplest form, which seems to be a good approximation, is a Gaussian distribution with a width σ [49]. In this approximation, the intensities of the (00n) reflections from the smectic layers are reduced by a factor $\exp(-\sigma^2 n^2 q_0^2)$ analogous to the Debye-Waller factor of crystals, and the intensity ratio of the second to the first harmonic is given by

$$I_{002}/I_{001} = F^2(2q_0)\, \exp\left(-3\sigma^2 q_0^2\right)/2F^2(q_0). \tag{6}$$

In accordance with Eqs. (4) and (6) the translational order parameters may be written as $\tau_n = \exp(-\sigma^2 n^2 q_0^2/2)$. For typical values of parameters cited in Eq. (6), $I_{002}/I_{001} \approx 10^{-3}-10^{-4}$, $F^2(2q_0)/F^2(q_0) \approx 10^{-1}$ and $q_0 \approx 0.2$ Å^{-1}, the corresponding width $\sigma \approx 5$–6 Å and the ratio $\tau_2/\tau_1 \approx 0.1$. These estimates indicate that low molecular mass thermotropic smectics are, as a rule, essentially "soft" layered systems with a nearly sinusoidal density wave $\rho(z)$. However, for some particular types of layering (for example, for perfluorinated smectics) the ratio τ_2/τ_1 is much higher, indicating deviations of the density distribution function from a pure sine form (see Sect 4.1).

The infra-red (IR) dichroism is a powerful tool for determination of orientational order parameters for various molecular fragments and evaluation of the molecular conformations in different types of mesophases [50]. As a rule, either a biphenyl moiety or a nitrile group are used as labels to probe the orientational order. For example, the nitrile group at the end of hydrocarbon chain reveals the "melting" of the latter on transition from nematic to smectic A_d phase [51]. However, the scope of the method is much broader. One can measure order parameters for different molecular fragments and determine the preferable molecular conformation in LC phases. In combination with the X-ray diffraction it makes possible to establish the precise picture of molecular packing. Recently an IR dichroism technique has been applied to the determination of the tilt angle in the smectic C phase [52] and for studying conformations of polyphilic mesogens [44, 45, 53].

The classical scheme for dichroism measurements implies measuring absorbances (optical densities) for light electric vector parallel and perpendicular to the orientation of director of a planarly oriented nematic or smectic sample. This approach requires high quality polarizers and planarly oriented samples. The alternative technique [50, 53] utilizes a comparison of the absorbance in the isotropic phase (D_i) with that of a homeotropically oriented smectic phase (D_h). In this case, the apparent "order parameter" for each vibrational oscillator of interest S^* (related to a certain molecular fragment) may be calculated as $S^* = 1-(D_h/D_i)\cdot(1/f)$, where f is the thermal correction factor. The angles of orientation of vibrational oscillators (ϕ) with respect to the normal to the smectic layers may be determined according to the equation

$$S^* = SS_\phi = S[(1 - (3/2) \sin^2 \phi] \tag{7}$$

where S is the true orientational order parameter of the smectic phase. Hence, in the smectic A phase, S can be conveniently measured by recording the dichroic ratios for either parallel ($\phi = 0$) or perpendicular ($\phi = \pi/2$) bands. For complicated and flexible molecules, S and S^* are not expected to have the same values for all oscillators. For example, in order to describe the molecular conformations for the polyphilic mesogens, the order parameters and the angles of orientation of the main molecular fragments: the biphenyl core, the hydrocarbon chain and the perfluoroalkyl tail have to be determined from the dichroism of the number of characteristic molecular bands [44, 45, 53].

In the smectic C phase, two angles are necessary to describe the position of a molecule or molecular fragment with respect to the layer frame. Two other angles are needed to locate a given oscillator with respect to the molecular frame. However, in the homeotropic geometry, using unpolarized light and an azimuthally disordered texture, the apparent order parameter can be written in a simple form, similar to Eq. (7), both for the longitudinal and transverse oscillators [45, 53]. Knowing the value of the true orientational order parameter S in the smectic C phase, it is therefore possible to determine the tilt angle of a certain molecular fragment from dichroic data.

3
Polar and Steric Frustrations

Frustration has emerged as a key concept in describing the variety of particular smectic phases and re-entrant phenomena in liquid crystals. Considering the smectic layers formed by asymmetric molecules, we need not only to specify a density modulation to define the existence of layers, but also the positions of polar or steric dipoles in the molecules with respect to the layers [54] (Fig. 4). This assumes the existence of two or more competing interactions favouring different periodicities. In cases where the energies of the competing interactions are close to one another, different local orientational configurations equally minimize the energy. This is a typical situation for which the frustrated states occur.

The frustration effects are implicit in many physical systems, as different as spin glass magnets, adsorbed monomolecular films and liquid crystals [32, 54, 55]. In the case of polar mesogens the dipolar frustrations may be modelled by a spin system on a triangular lattice (Fig. 5). The corresponding Hamiltonian consists of a two particle dipolar potential that has competing parallel dipole and antiparallel dipole interactions [32]. The system is analyzed in terms of dimers and trimers of dipoles. When the dipolar forces between two of them cancel, the third dipole experiences no overall interaction. It is free to permeate out of the layer, thus "frustrating" smectic order.

The competition between the polar and steric dipoles of molecules may also lead to internal frustration. In this case, the local energetically ideal configuration cannot be extended to the whole space, but tends to be accomodated by the appearance of a periodic array of defects. For example, the presence of the strong steric dipole at the head of a molecule forming bilayers will induce local curvature. As the size of the curved areas increases, an increase in the corresponding elastic energy makes energetically preferable the

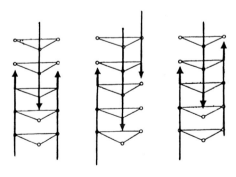

Fig. 5. Nearest neighbour triplet dipole configurations on triangular lattice, which play an important role in stabilization of different types of smectic A packings

transition from uniform bilayers to in-plane domains with opposite signs of local curvature, separated by antiphase borders (defect walls) [56]. The resulting structure may be one dimensional (modulated), two-dimensional (of the type of smectic $\tilde{A}$ or $\tilde{C}$ phases, Fig. 4), or 2D soliton-like [54].

The picture of dipolar frustrations depicted in Fig. 5 is of course an idealized one. Evidently the antiparallel dipolar configurations occur at the expense of the packing, which is less favourable than in the monolayer or bilayer phases shown in Fig. 3c. The cross-sections of the central parts of associated molecules exceeds that of alkyl chains due to the acoplanarity of phenyl rings and the planes of bridge fragments. This leads to an unfavourable decrease in the entropy of packing and restricts the discrete set of points at which the mesogenic molecules can overlap with one another. The unfavourable packing of the central cores may be partially relieved by the tail chain flexibility of the molecules. Thus the packing entropy effects, molecular corrugation and tail chain flexibility have to be included in a realistic theory of frustrated smectics.

3.1
Smectic Layering in Terminally Polar Mesogens

A number of experiments performed since the late 1970s [11–13, 30] found a variety of smectic phases with a broken head-to-tail symmetry among mesogens possessing strongly polar cyano or nitro terminal groups (Fig. 6).

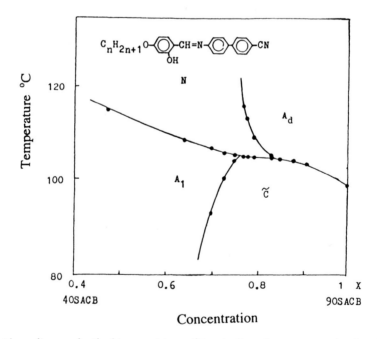

Fig. 6. Phase diagram for the binary mixture of terminally polar mesogens showing the re-entrant nematic (N), monolayer (A_1), partially bilayer (A_d) and two-dimensional ($\tilde{C}$) phases (Lobko et al. [47])

Some of these phases occur in reentrant phase diagrams showing single [30] and sometimes multiple reentrant behaviour, described by the phase sequence I-N-A_d-N-A_d-N-A_1-$\tilde{C}$-A_2-C_2 on cooling [46, 57].

The smectic phases A_1, A_2 and A_d have the same macroscopic symmetry, differing from each other in the wavelength of spacing. Hence it is possible to go from A_1 to A_d or from A_d to A_2 by varying only the layer periodicity in a continuous or discontinuous way(with the jump in the layer spacing d). Smectic–smectic transition lines of first order may terminate at a critical point, where the differences between the periodicities of the smectic A phases vanish, providing a continuous evolution from an A_d to bilayer A_2 phase [12].

Prost et al. [54] developed a Landau model to describe these phases and the features of the phase diagrams of polar mesogens.Their theory utilizes two order parameters: the conventional mass density $\rho(\mathbf{r})$ and a polarization wave $P(\mathbf{r})$, which describes long-range head-to-tail correlations of asymmetric molecules. The theory determines the regions of stability of all of the phases described above and provide an explanation for single re-entrance in the phase diagrams. Multiple re-entrance, however, requires consideration of a suitable microscopic model (see below). Consideration of layer displacement fluctuations was shown to be important near the A_1-A_d critical point. At this point the compressional elastic constant B of the smectic layers vanishes in analogy with the behaviour of the inverse compressibility at the liquid-vapour critical point [58]. Thus the system is nematic-like in the vicinity of the critical point and might be replaced by a "bubble" [59]. The region of the nematic phase might be expanded on the whole A_1-A_d transition line, thus inducing the gap between the regions of stability of these phases. The frustration, i.e. melting to the re-entrant nematic phase, might be escaped by the formation of "antiphases" $\tilde{A}$ and $\tilde{C}$, which effectively reduces the free energy of the system due to the coupling between the two order parameters of terminally polar smectics [54]. It is not surprising, therefore, that those phases occupy a position in phase diagrams intermediate between the regions of stability of A_1 and A_d phases (Fig. 6). Another way to escape frustration is to go over to an incommensurate smectic phase A_{inc} with two colinear modulations at wavevectors q_1 and q_2, the ratio q_1/q_2 being irrational [54]. There are serious doubts that the A_{inc} phase can exist in polar smectic mesogens with liquid layers [60], although incommensurate smectic E [61] and smectic B [62] phases have been observed.

3.2
Examples of Steric Frustrations

So far we have considered the diversity of the types of smectic layering as being due to dipolar origins. Now we discuss the particular types of packing arising from the asymmetry of molecular shape as well as from the tail chain flexibility. Pelzl et al. [39] observed re-entrant nematic behaviour in binary mixtures of mesogens without polar terminal groups. Instead, however, at least one of the components of the mixture was strongly sterically asymmetric. Evidently, the smectic A layering in such a system can occur only at the expense of packing which is less favourable than that in the nematic phase.

The particular types of smectic layering resulting from an extremely asymmetric molecular shape were found among the swallow-tailed compounds studied by Diele et al. [63]. The formation of the smectic phases for these compounds is unfavourable due to an excess free volume in the region of the central cores. However, the partially bilayer A_d phase may be stabilized in a certain temperature-concentration range by the incorporation of shorter rod-like molecules into the free space between the bulky swallow-tailed ends of the molecules (the so-called filled smectic A phase). The modulated smectic phase of the type $\tilde{A}$ was found also for a special case of so called tuning-fork-like molecules [40]. Different types of particular nematic (with strong A_d and tilted fluctuations) and smectic phases, where molecules are overlapped or form modulated structures, have also been reported for mesogenic molecules of complicated spatial configuration. These include laterally branched compounds [64], twin molecules [65] and dimeric mesogens in which mesogenic cores are linked through a flexible spacer [21].

The latter case represents a special interest because the incommensurate A_{inc} phase, which was expected to exist in polar mesogens [54], was surprisingly found by Hardouin et al. [42] in a dimesogenic compound without polar head. Rather, the molecule consists of two chemically different mesogenic units: a cholesteryl one and a classical aromatic core connected through an alkylene flexible spacer. The formation of the A_{inc} phase was detected by the occurrence of three colinear density modulations of incommensurate wavelengths. The largest spacing corresponds to the head-to-tail arrangement of the molecules as a whole, while the smallest periodicity (approximately 20 Å) corresponds either to the length of the cholesteryl moiety or to the length of the aromatic core. The A_{inc} phase was found only for the mesogen containing five carbon atoms in the alkylene spacer chain. Compounds with shorter and longer spacers show a rich polymorphism, but no sublayer segregation. It is likely that the presence of the bulky cholesteryl part and its non-aromatic nature combined with the planar shape favours the sublayer packing at a certain length of the flexible spacer.

The steric frustrations have also been detected in LC polymers [66–68]. For example, the smectic $\tilde{A}$ phase with a local two-dimensional lattice was found by Endres et al. [67] for combined main chain/side chain polymers containing no terminal dipoles, but with repeating units of laterally branched mesogens. A frustrated bilayer smectic phase was observed by Watanabe et al. [68] in main-chain polymers with two odd numbered spacers sufficiently differing in their length (Fig. 7).

Because of the odd number of conformational constraints, the direction of tilt of the LC groups regularly alternates from one spacer to another. The polymer chains form a bilayer structure, although another periodic density modulation in a direction parallel to the layer was also observed. In this picture, the bilayers are arranged in a small areas with the polymer chains in adjacent areas slide halfway along the layer normal (Fig. 7). From the examples presented above, we conclude that frustrated types of layering may be obtained without any dipolar interaction, but only by an appropriate design of the mesogenic molecules.

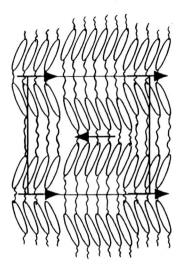

Fig. 7. Two-dimensional packing observed for the main-chain polymer with two odd numbered spacers of different length. *Solid lines* and *arrows* indicate the two-dimensional rectangular lattice and macroscopic polarization, respectively (Watanabe et al. [68])

3.3
Microscopic Aspects of Smectic A Layering

The molecular origin of particular types of smectic A layering has been considered by a number of authors [29, 32, 56, 69, 70]. The earlier theories are built upon the McMillan model of the nematic-smectic A phase transition. As we have discussed already, this model has serious flaws when trying to describe the phase behaviour of asymmetric mesogens. It is now clear that the restrictions of the McMillan model stem from neglecting the hard core repulsions between molecules and, consequently, the changes of the packing entropy. These variations may be extremely large for terminally polar molecules having a tendency for dipole–dipole association or for molecules of strongly asymmetric shape forming steric dimers.

There are some successful attempts to apply the McMillan model to terminally polar LCs forming dimers by inclusion packing entropy effects [56a, 69]. As noted by Longa and de Jeu [56a], an increase in the number of dimers in the system stabilises the A_d phase due to a more favourable ratio between the length of the central part of the associate r_0 and that of the whole dimer L_d, compared with monomers. This allows one, in the spirit of McMillan theory, to consider the nematic-smectic A_d transition as a percolation type transformation, occuring at a certain concentration of dimers [71]. The direct reason for the nematic re-entrant behaviour is an unfavourable decrease in the packing entropy with the increasing number of dimers in the system at lower temperatures. Any other mechanism resulting in the unfavourable packing of molecules or their moieties in smectic layers causes the same effect. The regions of stability of diverse smectic A phases were determined by considering the competition between dispersive and dipolar interactions, in conjunction with steric repulsions [69]. It was found from estimates of these interactions that the ferroelectric smectic A phase with non-zero macroscopic polarization is strongly destabilized almost independently of the position of the molecular dipoles. In contrast, the dipole repulsions become quite small in the bilayer A_2 phase, provided that the dipoles are located at the end of molecules. In combination with the optimal packing, this makes the A_2 phase, as well as the $\tilde{A}$ and $\tilde{C}$ phases with local bilayer ordering, the lowest temperature smectic phases, as observed experimentally.

A spin-gas microscopic theory has been pursued by Berker et al. [32] to explain the multiplicity of smectic ordering and the re-entrance phenomenon in strongly polar mesogens. They have used a model Hamiltonian of the form

$$V_{12} = r_{12}^{-3}[A(s_1s_2) - 3B(s_1r_{12})(s_2r_{12})]. \tag{8}$$

When $A = B$, the expression at Eq. (8) represents the usual interaction energy of permanent dipoles. When $A < B$, parallel orientations of dipoles are favoured and when $A > B$, the interaction energy has a minimum for antiparallel dipoles.

The dipoles, represented by s_j, are localized at the end of the molecules and occupy n consecutive sites along the layer normal in a three-dimensional stacked triangular lattice (Fig. 5). The segmentation into n units (n = 4 or 5) arises from the molecular corrugation – the nearest neighbour molecules have a set of energetically preferred positions relative to each other. The transition from smectic A_d to re-entrant nematic phase corresponds to the nearest neighbour configuration in which two molecules form an antiparallel dimer in the same notch of the lattice. In this case there is a near cancellation of interactions between the "dimer" and the third member of the triplet that is now free to permeate out of layer, thus disturbing the smectic order. On the other hand, triplets which are displaced relative to each other have a net dipole and so can stabilize a layered structure through competition between the local ferroelectric or antiferroelectric order.

More complicated phase diagrams with multiple re-entrance were obtained when Van der Waals interactions and librational permeation (i.e. possibility for dipoles to change their position about corrugation minima) have been included. The additional inclusion of the benzene ring steric hindrance yields the model for tilted ordering, including appearance of smectic A and C phases with in-plane domain formation, resembling that of $\tilde{A}$ and $\tilde{C}$ phases. The corresponding phase diagram is depicted in Fig. 8 and shows good agreement with the experimental observations (see Fig. 6).

Parallel with dipolar frustrations, the tail chain flexibility is of great significance for the stability of the particular smectic phases. The presence of molecules with different chain configurations (conformers) stabilizes the partially bilayer A_d phase due to the denser packing of molecules in the layers. In fact the smectic A_d phase only exists for relatively long tail chains, starting from six to eight methylene units [11–13]. The population of the *gauche* conformers is higher for mesogenic molecules with lengthy tails and, therefore, longer homologues can have a denser packing in the A_d phase than molecules

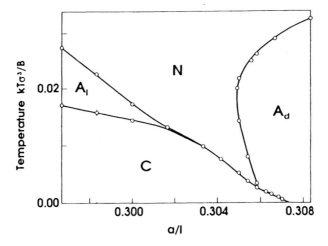

Fig. 8. Theoretical phase diagram for a moderate strength dipolar system (Berker et al. [32])

with the shorter tails. It is also symptomatic that the critical length of the tail chain which stabilizes the partially bilayer type of layering is of the same order of magnitude as the persistence length in linear hydrocarbon polymers [72].

With decreasing temperature, the tails become less flexible and this can induce either the transition to monolayer (bilayer) smectic phases or "melting" into the re-entrant nematic. A lattice model of smectics reproducing this behaviour was developed by Dowell [29]. It was found that the differences in the packing of the rigid cores and flexible tails-as a function of tail chain flexibility can stabilize smectic A_1 and partially bilayer A_d phases and even induce a transition to re-entrant nematic phase. Of course, the changes in the tail chain flexibility are not expected to be the universal mechanism of layer destruction. However, with strongly sterically asymmetric molecules, it may drive the re-entrant nematic transition as well as stabilize the certain types of layering for mesogens possessing large packing differences between the cores (or tails) of different nature. The examples of such behaviour were shown in Sect. 3.2.

4
Liquid Crystals with Incompatible Constituent Parts

The amphiphilic character of mesogenic molecules is widely accepted as playing an essential role in the stabilisation of smectic order [73]. Indeed, the central aromatic core and flexible aliphatic chains are different in chemical nature and, hence, incompatible with one another to some extent. This internal incompatibility was checked in experiments of Guillon et al. [73], who studied the miscibility of the constituent parts of mesogenic molecules, i.e. of a linear paraffin $C_{12}H_{26}$ and a rigid aromatic molecule comprising two phenyl rings. In fact, a large region of immiscibility in the liquid state was observed, while the molecule composed of these two units exhibits a smectic behaviour. This indicates that the smectic layering can be considered as a result of the incompatibility between the constituent parts of mesogenic molecules. Such behaviour is similar to the microphase separation observed in diblock copolymers [74] and diluted side chain mesogenic polymers [75].

In recent years more sophisticated mesogenic molecules with a sequence of chemical fragments of different nature and steric configuration have been synthesized [18–20, 41, 42, 76]. The main efforts have been applied for partial or total substitution of fluorine atoms for hydrogen atoms at one of the tail chains of a molecule. The incompatibility of perfluoro chains with aliphatic moieties is a well known fact [77]. Another example represents the so-called polyphilic mesogens, originally synthesized by Tournilhac and Simon [76b]. These molecules are four-block in nature and made up of the following sequence of chemical moieties: a rigid perfluoroalkyl chain, a flexible hydrocarbon chain, a rigid biaryl core and again a perfluoroalkyl chain (Fig. 9).

The basic idea that stimulated the synthesis of these compounds was that in polyphilic LCs three or more chemically different fragments could segregate in a special way, resulting in the formation of particular types of layering, including the polar packing with non-zero macroscopic polarization. Indeed, later experiments confirmed an existence of polar types of ordering in some polyphilic mesogens [78]. On theoretical side, Prost et al. [79] developed a Landau theory for the polar ordering of polyphilic mesogens. A rich sequence of phases showing polar, antiferroelectric and modulated packing has been found. Petschek et al. [80] using Monte Carlo computer simulations of three and five block molecules predicted effects of intralayer segregation and sequenced in polar order (flexible chain, short core, flexible chain, long core, flexible chain) smectic layers. In fact, effects of sublayer segregation were observed by Hardouin et al. [42] in three block mesogens and Lose et al. [41] in three block terminally fluorinated compounds. Below we will consider the peculiarities of smectic layering in LCs with chemically and (or) sterically incompatible constituent parts.

4.1
Terminally Fluorinated Mesogens

These compounds present a new class of LC molecules having a perfluoroalkyl tail on one end and an ordinary alkyl tail on the other. Earlier studies of several groups indicated that terminal fluorination considerably extends the temperature range of the smectic phase [81]. Moreover, mesogens incorporating

Fig. 9. Molecular model of polyphilic (four-block) mesogen

fluorine atoms show improved characteristics for electro-optic devices, including ferroelectric liquid crystal display applications, and their development is currently of great interest [82]. Recent investigations of perfluorinated mesogens have revealed a number of remarkable properties, as compared to the fully hydrogenated molecules. Sigaud et al. [77] found a phase separation associated with a consolute critical point in a smectic A mixture of a polar compound with an aliphatic tail and a nonpolar compound with a perfluoro-alkyl tail. Usually smectic A phases of different molecules are perfectly mixed [83]. Furthermore, fluorinated compounds exhibit layer-by-layer thinning transitions above the bulk smectic A-isotropic transition [84] and a chiral-symmetry-breaking twist-bend instability in achiral free-standing films [85].

The X-ray diffraction patterns of fluorinated compounds display up to four orders of reflections (00n) resulting from the smectic A layering [37, 38, 86, 87]. The intensity ratios of the second and third harmonic to the first – I_{002}/I_{001}, I_{003}/I_{001} – were found to be as large as 5–10%. This is in sharp contrast to the conventional hydrogenated smectics A where the ratio I_{002}/I_{001} is two to three orders of magnitude less [43]. Using a Gaussian approximation, the width σ of the molecular centre of mass distribution function within the layers and the ratio τ_2/τ_1 of the first two translational order parameters may be calculated from the intensity of the (00n) reflections (see Sect. 2.3). The values $\sigma \approx 3$–4 Å and $\tau_2/\tau_1 \approx 0.3$–$0.5$ for perfluorinated mesogens considerably deviate from that of the smectics with aliphatic tails: $\sigma \approx 5$–6 Å and $\tau_2/\tau_1 \approx 0.1$ [37]. This correlates with the considerably larger values of the compression elastic constant B in terminally polar mesogens as compared to hydrogenated LCs [86]. The smectic layers in perfluorinated mesogens thus appear to be much more well defined than in a conventional smectic A phase, and the corresponding one-dimensional density wave $\rho(z)$ is far from being purely of the sine form.

Now, what particular properties of perfluorinated chains exert their influence on the smectic ordering? Firstly, the Van der Waals radius of fluorine considerably exceeds the corresponding value for a hydrogen atom. Consequently, the cross section occupied by closely packed perfluoro chains ($s_F \approx 28$ Å^2) is about 30% larger than that of alkyl fragments ($s_H \approx 20$ Å^2) and of the cylindrically averaged central cores of molecules ($s_C \approx 22$ Å^2). Secondly, in accordance with quantum-chemical calculations by Koden et al. [81c], the activation energy of the *trans–gauche* and *gauche–gauche* rotations in fluorocarbons (decaflourobutane) is a factor of three to four higher than in linear hydrocarbons (butane). Thus the internal rotations of the perflorinated chain are fairly restricted and the number of *gauche* conformers is considerably reduced. Because of the steric hindrance of the consecutive CF$_2$ moieties, this results in a rigid rod-like structure, existing in a 15/7 helical conformation (as in polytetrafluoroethylene), while hydrogenated chains are in a planar zigzag configuration. The difference in the cross-sections of the constituent molecular fragments determines the strong wedge shape of the terminally perfluorinated molecules (Fig. 10a). Enhanced rigidity of the smectic layers formed by these molecules may be attributed to purely steric factors associated with the bulky shape and the conformational stiffness of the

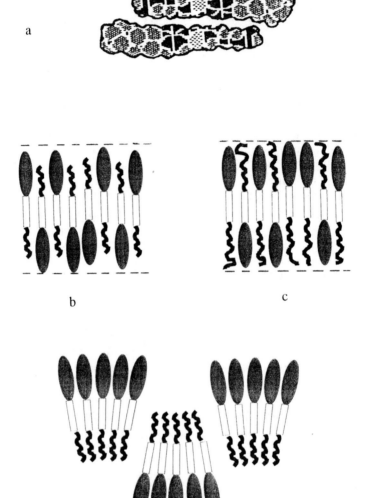

Fig. 10a–d. Various smectic packings for terminally fluorinated mesogens: **a** an antiparallel arrangement, the wedge shape of the molecules is clearly seen; **b** a dimer-like packing for the molecules in which $L_F > L_{H}$; **c** antiparallel packing for the case $L_F < L_{H}$, layer spacing is independent of the length of hydrocarbon tails; **d** a parallel packing of the wedge shaped molecules. Steric mismatch leads to splayed arrangment which may be relieved by the formation of the two-dimensional (modulated) structure. Rigid cores, fluorinated and hydrocarbon tails are represented by *rectangles, hatched ovals* and *zig-zag lines*, respectively

perfluorinated chains. Note that the presence of the flexible (hydrocarbon or semifluorinated) chain in terminally fluorinated mesogens is essential for the stability of liquid crystalline order [81, 88]. To the best of our knowledge, molecules with two fully fluorinated chains fail to exhibit mesomorphic states. Furthermore, $-CF_3$ and $-OCF_3$ molecular groups, as distinct from the $-CH_3$ fragments, possess a fairly large longitudinal dipole moment (approximately 2–3D [89]), which may favour dipolar frustrations.

Lobko et al. [37] and Rieker and Janulis [87] have used X-ray diffraction to determine the exact structure of the smectic A layers formed by perfluorinated molecules. Compounds with the short $-OCF_3$ moiety at the end of the molecule display the usual monolayer smectic A_1 phase [37]. Compounds with lengthy (three to seven $-CF_2-$ fragments) perfluorinated chains generally show smectic layering which is incommensurate with the molecular length. This may be equal, larger or even smaller than the molecular length L ($L = L_C + L_F + L_H$, where L_C is the size of the molecular rigid core) depending upon the ratio between the length of the fluorinated L_F and hydrogeneted L_H tails. This points to some type of dimeric organization of the molecules, which may be either parallel or antiparallel. Rieker and Janulis [87], in their study of homologous series of alkyl[(dihydroperfluoroalkoxy)phenyl] pyrimidines, have found that the smectic layer thickness depends only upon the length of the fluorocarbon tail and is independent of the length of the hydrocarbon tail. This indicates that the molecules are organized into antiparallel pairs, i.e. with fluorinated and hydrogenated moieties adjacent, such that the hydrocarbon tails do not contribute to the smectic layer spacing (Fig. 10b,c).

If $L_F \approx L_H$, the monolayer smectic A_1 phase is formed. In the case that $L_F > L_H$, the layer spacing d exceeds the molecular length L, which corresponds to the formation of the smectic A_d phase (Fig. 11a,b). In the opposite case, $L_F < L_H$, the layer spacing is less than L, and the flexible hydrocarbon chains just fill the space which is determined by the fluorinated tails (Fig. 10c).

A similar type of dimeric layer organization was eventually observed by Diele et al. [38] for the homologous series of perfluoralkylethyl(4-phenylbenzylidene)imines, in which the biphenyl fragment is linked via a short spacer to a fluoroalkyl chain (Fig. 11a). These molecules have no hydrocarbon tail at all. The layer spacing d in the smectic A phase was found to be much larger than the molecular length L, while the difference $\Delta = d-L$ increases with the length of the perfluorinated chain. This corresponds to the formation of the smectic A_d phase which is built up of antiparallel steric dimers (Fig. 11a).

A model with overlapping perfluoroalkyl tail should be excluded, since in this case the difference Δ is independent of the length of the fluorinated chain. The calculations for the molecular form factor gives a reasonable agreement with the intensities of successive (00n) harmonics for the model with overlapping aromatic parts of the molecules and the tilt (approximately 35°) of perfloro chains [41c]. This model also satisfies the requirements for dense filling of space. The smectic layers in the dimeric smectic phase are well defined ($\sigma \approx 2.5-3$ Å) and consist of two sublayers of the fluorinated and aromatic parts of the molecules.

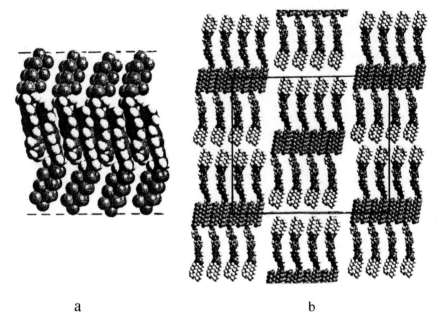

a   b

Fig. 11a,b. Examples of one and two-dimensional structures for terminally fluorinated mesogens: **a** antiparallel dimeric layering for the compounds with one terminal fluorinated tail (Diele et al. [38]); **b** two-dimensional (columnar) packing observed for the swallow-tailed fluorinated compounds; solid lines indicate two-dimesional rectangular lattice (Lose et al. [41])

A denser packing in the smectic A layers is achieved for an antiparallel arrangement of the wedge shaped perfluorinated molecules (Fig. 10a–c). Parallel packing would produce higher-energy states due to the local splay of neighbouring molecules (Fig. 10d). This is confirmed independently by surface tension measurements on free-standing films of perflorinated smectics [90]. Thus we conclude that for terminally fluorinated mesogens steric interactions dominate the polyphilic interactions which tend to segregate the fluorocarbon and hydrocarbon tails. These observations markedly contradict the behaviour of semifluorinated alkanes – $F(CF_2)_m(CH_2)_nH$ (m and n can vary from 10 to 20) – which is governed by their amphiphilic character [91]. Antiparallel arrangement is characteristic only of the high temperature lamellar crystalline phase, while the low temperature lamellar phases display double-layered structures with segregated sublayers. The differences in the packings of the perfluorinated mesogens and of the semifluorinated alkanes most likely are due to the presence of the highly polarizable aromatic cores in LC molecules. However, for more sophisticated perfluorinated molecules, such as the swallow-tailed compounds, the effects of intralayer segregation have been observed [41a]. The smectic A layering for these mesogens (with the fluorinated part placed either in the branched part of the molecule or in the unbranched one) is generally characterized by an antiparallel packing of the molecules. However, on cooling, the smectic A phase is replaced by a less symmetrical columnar phase with a two-dimensional rectangular lattice (Fig. 11b). This transformation is indicated by the appearance of the sharp equatorial (0k0) and combined (0kn) reflections on the X-ray diffraction patterns in addition to meridional (00n) reflections. The longitudinal parameter a of the two-dimensional lattice corresponds to the length of a dimer in which the molecules are shifted so that the perfluorinated parts are adjacent. The transverse parameter b determines the characteristic size of the segregated area of bilayers. In this structure the incompatible parts of the molecules are separated on the small scale of the order of $b/2 \approx 90$ Å, while on the larger scale the system remains in the mixed state. Thus, while in the high temperature smectic A phase the packing is determined by steric interaction of the wedge shaped molecules, the increasing tendency for the segregation of incompatible moieties on cooling leads to the formation of the phase with rectangular lattice where the requirements of segregation are locally satisfied.

In closing this section, we should mention that the remarkable properties of perfluorinated chains in mesogenic molecules have rich potential applications in different areas. For example, these materials show an extremely large negative thermal expansion coefficient in smectic layer spacing [87]. Such thermal behaviour originates from the relatively easy sliding of the antiparallel wedge-shaped molecules and is important for creating mixtures for manufacturing defect-free surface-stabilized ferroelectric LC cells for electro-optical devices [82]. Because fluorocarbon chains are both hydrophobic and oleophobic, and have very low surface tension, these materials are intensively used in such applications as electroplating baths, dispersants for herbicides, repellents for both water and oil, etc [92]. Further studies of perfluorinated molecules with an intrinsic tendency for self organization will shed new light on the fundamental properties of fluorocarbon chains.

4.2
Polyphilic Liquid Crystals – Structure and Phase Behaviour

Specific properties and motivation for synthesis of the polyphilic mesogens have been discussed in the papers of Tournilhac et al. [76]. The structure and phase behaviour of two types of polyphilic molecules have been studied later by means of X-ray diffraction and IR dichroism technique [44, 45]. The first type of mesogens, with general formula $F(CF_2)_8(CH_2)_{11}O$—ϕ – ϕ—COO-CH_2CF_3 (where ϕ stands for the phenyl ring), consists of perfluorinated chains of different length at the ends of the molecule (Fig. 9). Compounds of this kind are abbreviated as I and II and differ from each other by the position of the ether and ester fragments relative to the biphenyl core. The another type of polyphilic compounds have a strongly polar cyano group at one end of the molecule: $F(CF_2)_n(CH_2)_mO$—ϕ – ϕ—CN and are abbreviated as F_nH_mOCB [45].

Both compound I and a mixture of the two derivatives M70 (70% I:30% II) display the usual monolayer smectic A_1 phase on cooling down from the isotropic liquid. At lower temperatures the particular smectic phases (X and X') showing polar behaviour have been observed [78]. The smectic X' phase in the mixture M70 displays three orders of X-ray diffraction from a lamellar structure with the spacing $d \approx 35$ Å. This value is significantly less than the molecular length $L \approx 43$ Å. This corresponds to a tilted arrangement of molecules or their fragments within the smectic layers. The (00n) diffraction peaks are not resolution limited – the correlation length in the direction parallel to the layer normal $\xi_\parallel$ is about 200–300 Å. Thus, the lamellar structure in the smectic X′ phase appears to be strongly defective with regions of the smectic C layering which are not correlated beyond about 10 smectic spacings.

The IR dichroism measurements allowed a fairly precise determination of the preferential molecular conformations both in the smectic A_1 and X′ phases (see Sect. 2.3). In the smectic A_1 phase the biphenyl moiety is parallel on average to the layer normal, while the hydrocarbon and perfluorinated fragments are tilted at angles 18 and 32°, respectively. The phase transition to the smectic X′ phase is accompanied by a dramatic change in the main molecular conformation – now all the fragments are strongly tilted with respect to the layer normal (especially the biphenyl core which tilts at an angle of around 56°) (Fig. 12).

The smectic X phase in the pure compound I shows additionally to the smectic C layering the two-dimensional (modulated) structure in which the smectic C layers are periodically shifted with a respect to one another by half a layer spacing.

X-ray diffraction reveals a strong change of the character of positional order in the plane of the smectic layers for the polyphilic LCs. The conventional smectics usually show a broad, liquid-like peak centered at $q_\perp \approx 1.4$ Å^{-1}, corresponding to an average intermolecular distance of 4.5–5 Å and an in-plane correlation length $\xi_\perp \approx 10$–15 Å. The situation is quite different in the case of polyphilic compounds where the perfluorinated fragment is effectively decoupled from the rigid core via a flexible hydrocarbon spacer. The in-plane scattering in the smectic X and X′ phases displays three distinct peaks with the intermolecular distances $d_1 = 5.2$ Å, $d_2 = 4.8$ Å and $d_3 = 4.4$ Å (Fig. 13).

The first peak most likely arises from short-range correlations ($\xi_\perp \approx 3$–4 molecules) of the rotationally averaged molecules as a whole, while the second and third peaks, which dominate the intensity profile, show a dramatic increase in the intra-layer positional correlations. The correlation lengths $\xi_{\perp 2}$ and $\xi_{\perp 3}$ reach a value ≈ 35–40 Å, that is, approximately eight molecules are involved in the local in-plane order. The distance 4.8 Å is typical for close packing of the fluorinated chains [93], while the period $d_3 = 4.4$ Å might be related to the closely packed disordered hydrocarbon chains = The two types of preferential nearest neighbour contacts observed in smectic X and X′ phases point to segregation of the perfluorinated and hydrocarbon moieties. The differences in packing of various fragments along the molecular chain have been observed earlier in molecular dynamic simulations of monolayers of semifluorinated amphiphilic molecules [93b]. The bulky tilted fluorinated tails are close packed into distorted hexagonal structure, leaving considerable free volume at the positions of CH_2 groups. As a result, the concentration of *gauche* conformers for methylene fragments becomes an order of magnitude larger than for a fully fluorinated tail. The flexibility so generated leads to the

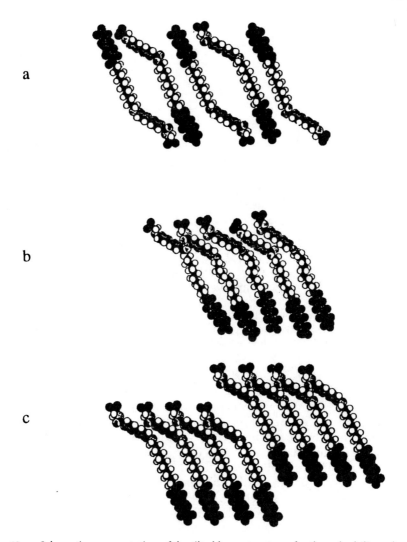

a

b

c

Fig. 12a–c. Schematic representation of the tilted layer structures for the polyphilic molecules in a strongly fractured conformation: **a** the random up-down configuration; **b** polar packing of molecules within the layer; **c** two-dimensional (modulated) polar structure (Blinov et al. [44])

appearance of the additional diffraction peaks corresponding to packing of the hydrocarbon moieties. A similar behaviour, when two periodicities are not related to lattice planes, has been observed for other systems with flexible, almost molten hydrocarbon chains, for example in discotic mesophases where two diffuse peaks correspond to the distances between rigid cores and molten hydrocarbon chains [94] and in side-chain polymers where mesogenic moieties are decoupled from the main chain [22, 66].

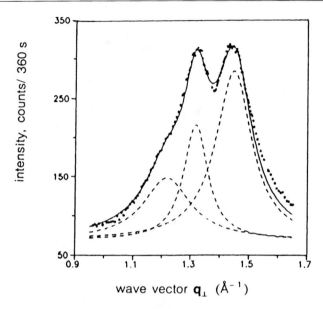

Fig. 13. X-ray intensity profile with diffraction vector along $q_\perp$ (in the plane of layers) in the smectic X' phase for the mixture of polyphilic compounds M70. The *solid line* through the data points is the sum of three Lorentzian peaks which are shown by *broken curves* (Blinov et al. [44])

Considering the smectic C layers of the strongly fractured polyphilic molecules in the random up-down configuration (Fig. 12), it is clear that this arrangement is not optimum from a purely steric point of view, due to an excess free volume in the region of differently tilted molecular fragments. The rigidity of the perfluorinated tails makes the compensation of the unfavourable molecular packing impossible as it usually occurs in the smectic phases formed by molecules with flexible hydrocarbon tails. Obviously, the low-energy structure of the smectic C layers, consistent with the fractured conformation of the polyphilic molecule and known interlayer spacing, is of the type shown in Fig. 12b. Now the projections of the biphenyl core and perfluorinated chain cross-sections onto the smectic layer plane are close to one another (≈ 35 Å^2) and the parallel (polar) packing of molecules is preferable, due both to steric and polyphilic interactions. If the polar direction is conserved in neighbouring layers, the resulting macroscopic structure is polar. In the opposite case of reversing the polar direction, the antiferroelectric structure will appear, which is characterized by doubling the layer periodicity. In the first experiments on polyphilic compounds only the polar type of layering has been observed. However, later it was found that, on slow cooling of the samples prepared as free standing films, the antiferroelectric structure may be also stabilized [95]. Note that for the polar structure of Fig. 12b, the slight difference in projection area for the biphenyl and the fluorinated chain results in a stress accumulated along a smectic layer. To compensate for the stress, the modulated structure of the type observed in compound I may be formed (Fig. 12c). Alternatively, the direction of the molecular tilt might alternate with a flip-flop of the polar structure of the layer. Thus, a number of defects (walls) separating domains of opposite polarity might be created, which explains the excess defects of the layer structure observed. In principle, the defects may form a regular superstructure resembling that of the so-called Twist Grain Boundary (TGB) phase in chiral A^* systems [96]. However, the size of the domains found for polyphilics is much smaller than the pitch of the A^* phase. This is not surprising, since the steric and polyphilic effects are much stronger than the influence of molecular chirality.

Now we will touch upon the layer structures of homologue series of polyphilic mesogens comprising the strongly polar end group F_nH_mOCB [45]. In contrast to their fully hydrogenated counterparts, the F_nH_mOCB compounds show a very broad range (up to 100 °C) of partially bilayer smectic A_d phase (see also [81d]). The same was also observed for cyanobiphenyls bearing a bulky pentamethyldisiloxane tails [97]. In addition to A_d phase, the tilted dimeric C_d phase has been detected over a wide temperature range for the compounds possessing relatively long alkyl and perfluoroalkyl chains $(n,m \geq 5)$ (Fig. 14).

The layer spacing for smectic A_d and C_d phases significantly exceeds the molecular length $(d/L = 1.35-1.65)$ and increases as the $A_d \Leftrightarrow C_d$ transition is approached from above. However, the temperature variations of the layer spacing in the C_d phase are very small in comparison to the classical smectic A-smectic C transition (Fig. 14). This unique behaviour results from the fact that both the tilt angles for different molecular moieties and the relative

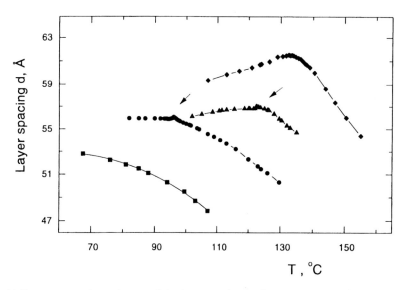

Fig. 14. Temperature dependences of the layer spacing in the smectic A_d and C_d phases for terminally polar polyphilic compounds: $F_4H_{11}OCB$ (*squares*), $F_6H_{11}OCB$ (*circles*), $F_8H_{10}OCB$ (*triangles*) and $F_{10}H_{11}OCB$ (*diamonds*). The *arrows* indicate the smectic A_d – smectic C_d phase transition points (Ostrovskii et al. [45])

displacement of molecules in dimers contribute to the layer spacing. For example, for $F_6H_{11}OCB$ the average tilt angle of the perfluoroalkyl moiety θ_F determined from the IR dichroism measurements is about 30° in the smectic A_d phase and gradually decreases when approaching the smectic C_d phase (Fig. 15). The average tilt angle of the alkyl fragment θ_H behaves in a similar way, attaining the small value of about 4–5° at the phase transition point. The cyanobiphenyl core is parallel on average to the layer normal in the A_d phase and tilts continuously to values above 16° in the C_d phase. Bartolino et al. [98] were the first to show that an aromatic rigid core tilts somewhat independently of the long aliphatic end chains within the smectic C layers (the so-called "zig-zag" model). This was confirmed later by precise X-ray and optical measurements of the smectic C phase for a number of mesogens [43b, 99]. The tilted smectics formed by polyphilic molecules show a much more pronounced "zig-zag" configuration, which determines the unique properties of these materials.

The basic structure of the smectic A_d phase with strongly overlapped central cores has the same origin (i.e. strong dipole–dipole correlations) as for a large variety of terminally polar mesogens (see Sect. 3.1). However, the fluorinated tails are much bulker than the aliphatic and aromatic fragments: $\sigma_F = s_F/\cos\theta_F$ is about 32–35 Å^2. At relatively high temperatures the presence of *gauche* bonds in the alkyl chains ensures the large tilt angles of the perfluorinated moieties. This makes it possible to fill the free space in the regions outside the overlapped rigid cores in a dense way, thus effectively stabilising the smectic

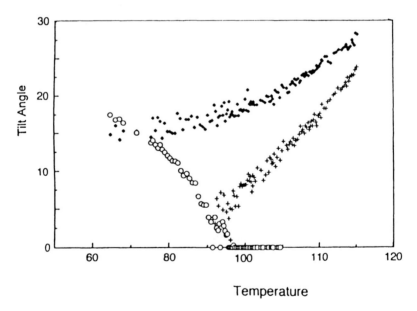

Fig. 15. Tilt angles of the different molecular fragments as a function of temperature for polyphilic compound $F_6H_{11}OCB$: aromatic core (*circles*), alkyl chain (*crosses*) and perfluoroalkyl chain (*diamonds*) (Ostrovskii et al. [45])

A_d layering (Fig. 16a). On the other hand, additional conformational entropy related to the sequenced hydrocarbon-perfluorinated chain prevents the crystallization of the aromatic parts of the molecules in the smectic layers as occurs for hydrocarbon analogues of the same tail length [4]. The stability of C_d phase in polyphilics is more delicate in nature and will be discussed below.

4.3
Tilted Smectic Phases

Discussions on the molecular origin of the tilted smectic phases go back for more than two decades. A number of microscopic models have been proposed [32, 100, 101], but no consensus has been reached. We note that the smectic C structure is unfavourable from the packing entropy point of view [100b]. Tilted molecules occupy more area in the smectic plane and, therefore, in the smectic C phase there is more excluded space than in the orthogonal phases. As a result, the packing entropy is decreased. Thus, additional interactions favouring the molecular tilt are required to stabilize the smectic C phase. These interactions may be of electrostatic [100] or steric nature, resulting from the anisotropic shape of mesogenic molecules [32, 101]. Some theories (for example the dipole model of McMillan [100a] or steric model of Wulf [101a]) imply that molecular rotation around the long axis is essentially frozen in the smectic C phase. This is in contradiction with experimental data [102]. At the same time, in the theory of Van der Meer and Vertogen [100b], the molecular tilt is caused by the induction interaction between the off-centre transverse dipoles and the polarizable core of a neighbouring molecule. This induction interaction is quadratic in the dipole moment and therefore compatible with slightly hindered rotations of molecules in the smectic C phase. Indeed, experimental studies reveal the importance of transverse dipoles for the stability of smectic C phase [103]. At the same time, Barbero and Durand [104] have shown, using macroscopic arguments, that the molecular tilt is an intrinsic property of any layered quadrupolar structure. In fact the effect of the induced dipole interaction is rather similar to that of point electric quadrupoles placed at the molecular centres [100b]. This means that

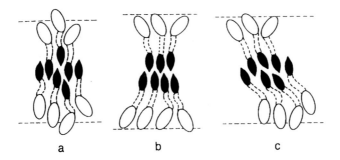

a   b   c

Fig. 16a–c. Structure of the smectic A_d and C_d layers for the polyphilic compound $F_6H_{11}OCB$: **a** smectic A_d layers; **b** steric mismatch in the limit case of bilayer A_2 layering; **c** smectic C_d layers

quadrupolar-type potential appears to be an universal potential to stabilize the tilted phases [105].

Thus the formation of tilted analogues of the smectic A phases, i.e. monolayer C_1 and bilayer C_2, is possible for mesogens with relatively large electric quadrupoles. In the case of strongly sterically asymmetric molecules (e.g., zig-zag shaped or dumbell shaped molecules, Fig. 3b) these quadrupolar interactions may be steric in origin. From this point of view observation of molecular tilt in the molecular dynamics simulations for a one-layer film of DOBAMBC in the absence of electrostatic interactions is not so surprising [106].

The stable dimeric C_d phase is hard to observe in conventional terminally polar mesogens [107]. This is a result of the unfavourable low values of packing entropy in the dimeric C_d phase as compared to smectic C_1 and C_2 phases. The situation is quite different in the case of terminally polar polyphilic molecules, where the rigid perfluoroalkyl and aromatic fragments at the ends of the flexible hydrocarbon chain may tilt (more or less independently) to compensate for the unfavourable decrease of the entropy of packing within the smectic layers. In the smectic A_d phase of F_nH_mOCB molecules, steric and dipolar interactions dominate the polyphilic interactions (Fig. 16a). As a result, moieties of different chemical nature are forced to coexist in the adjacent regions. The natural evolution of such a system with decreasing temperature would be the formation of a bilayer A_2 phase with segregation of the fluorocarbon, hydrocarbon and cyanobiphenyl moieties. Such a tendency is indeed reflected by the thermal behaviour of the layer spacing in the A_d phase (Fig. 14). However, in the extreme case of the bilayer A_2 phase, the steric mismatch between the perfluoroalkyl chains and cyanobiphenyl cores has too high an entropy cost for the minimization of the dipolar and polyphilic interactions (Fig. 16b). At this point the system has to find another solution, and transition to tilted smectic phase represents one of the possible compromises. Indeed, at the formation of the C_d phase, the molecules accept a strong zig-zag shape due to the inclination of the central core. This reduces the packing differences between the central part of dimers and terminal fluorinated chains and effectively stabilizes the dimeric type of layering (Fig. 16c).

Thus we conclude that the peculiarities of smectic layering for terminally polar polyphilic mesogens are determined by the conformational freedom of molecules mainly arranged in dimers as well as by the steric and chemical differences between different molecular fragments. The polyphilic liquid crystals show a unique mechanism for the compensation of the changes of packing entropy within the layers, related to the tilt of the different molecular fragments through differing angles. This effectively stabilizes the various types of dimeric and tilted phases formed by these materials.

5
Achiral Polar Phases and Chevron Structures

In recent years considerable efforts have been made to design achiral mesogenic molecules with a intrinsic tendency for polar self-assembly

[41b, 76, 108, 109]. Ferroelectric LCs have found extensive usage in both flat panel display applications and fast switching electro-optic devices [110]. Untill recently only chiral smectic C mesophases (C*) have been found to be ferroelectric [111] or antiferroelectric [112]. The polar natures of these materials are due to their chirality – in the smectic C* phase chirality breaks reflection symmetry in the plane containing the long molecular axis and layer normals (the tilt plane). This leads to a macroscopic polarization in the direction perpendicular to the tilt plane. There are, however, no fundamental reasons why the polar phases of achiral molecules should not exist. In this case the polarization along the average molecular long axis contributes to the macroscopic polarization. Such a LC would be a longitudinal ferroelectric with properties more similar to the conventional crystalline ferroelectrics than to a chiral C* phase. Such behaviour is generally not observed in LCs due to the head-to-tail orientational disorder and the consequent averaging effects. Even molecules possessing strong dipoles (for example, the cyanobiphenyls) have a tendency to form layered structures with an antiparallel orientation of the dipoles (smectic A_d or bilayer A_2 phases) where the average dipole moment is cancelled [113] (Fig. 4). The possibility for spontaneous breaking of this quadrupolar symmetry and thereby the manifestation of longitudinal ferro-electricity for achiral mesophases has been extensively debated over the last decade [79, 80, 114]. Computer simulations for spherocylinders with dipoles clearly indicate that, for the stable smectic phases, the layers are unpolarized almost independently of the location and magnitude of the molecular dipoles [115]. This means that additional chemical and (or) steric forces are required to overcome the unfavourable parallel arrangement of the dipoles.

The first experiments on polyphilic mesogens [78] manifest rather small values for the longitudinal polarization (≈ 10 nC cm^{-2}) of the smectic C phase formed by strongly fractured molecules. Further progress with these materials may be achieved if well oriented samples can be prepared, which is a non-trivial problem. Another example of the polar ordering in an achiral system are the so-called bowlic mesogens. Due to the extremely assymetric molecular shape, they are able to form columnar hexagonal phases with the polar ordering in the direction along the bowl axis [114a]. The similar behaviour was also predicted for columnar phases formed by dipolar discotics [114d,e]. Relatively large magnitude of polarization (of the order of 100 nC cm^{-2}) have been reported recently by Swager et al. [108] for certain discotic (bowl-like) metallomesogens. Ferroelectric behaviour has also been reported for achiral molecules with strongly bent cores (the so-called banana shaped molecules) forming smectic phases with liquid layers [109]. It was assumed that the steric dipoles of molecules are oriented in the same direction within the certain domains, which corresponds to the formation of a smectic phase with polar in-plane direction. However, as was shown recently by Link et al. [116], these achiral molecules actually spontaneously form chiral domains of opposite handedness. The thermodynamically stable state appears to be an antiferro-electric-racemic with the layer polar direction and handedness alternating in sign from layer to layer. An electric field induces the transition from the ground antiferroelectric state to the ferroelectric state.

Another interesting possibility to observe the effects of longitudinal pyro-, ferro- or antiferroelectricity in an achiral mesogenic system is based on the remarkable symmetry properties of smectics with alternating layer to layer tilt [117]. Such smectic phases (also called as chevron structures) have been known to exist for a long time [118]; however the enhanced attention given to them was stimulated by the observation of antiferroelectric behaviour for alternating tilt structures formed by chiral molecules [112]. The additional intermolecular interaction stabilizing chevron phases in comparison to the smectic C phase is the antiparallel pairing of the transverse dipoles of terminal chains, combined with steric repulsions of the molecular fragments in adjacent layers [102, 119]. The smectic C phase with alternating layer to layer tilt in achiral systems is generally nonpolar due to the up-down symmetry in the orientation of long molecular axes. However in the presence of the chain backbone for side chain mesogenic polymers, this head-to-tail inversion symmetry will be broken, resulting in an appearance of a two-fold axis in the tilt plane, along which the macroscopic polarization might appear (symmetry group C_{2v}, or *mm2* in crystallographic notation) (Fig. 17b). The first unambiguous confirmation of the polar effects in such a system was given recently by the experiments of Soto Bustamante et al. [120] for achiral polymer-monomer mixtures. Using X-ray analysis the mesophase was shown to be a bilayered smectic C with liquid-like layers. The mixtures show double P-E hysteresis loops measured by the pyroelectric technique and piezoelectric response. The ground state of the bilayered tilted structure has an alternating tilt with an in-plane antiferroelectric order (Fig. 17c) which can be switched to a ferroelectric state. The magnitude of the macroscopic polarization was found to be as high as 300 nC cm^{-2}.

We note that the bilayer smectic phase which may be formed in main-chain polymers with two odd numbered spacers of different length (Fig. 7), should also be polar even in an achiral system [68]. This bilayer structure belongs to the same polar symmetry group *mm2* as the chevron structure depicted in Fig. 17b, and macroscopic polarization might exist in the tilt direction of molecules in the layer. From this point of view, the formation of two-dimensional structure of the type shown in Fig. 7, where the polarization directions in neighbouring areas have opposite signs, is a unique example of a two dimensional antiferroelectric structure.

Finally, we should mention that the asymmetry of molecular shape, polyphilic effects and conformational constraints are the dominant factors in the stabilization of polar ordering in achiral mesogens. The examples presented above are, therefore, highly significant. They show that many liquid crystalline structures are intrinsically polar and may be effectively stabilized by suitable design of the mesogenic molecules.

6
Bridging the Gap Between the Smectics and Columnar Phases

Since the discovery of discotic liquid crystals [121], the mesophases formed by rod-like and disc-like molecules have been considered as belonging to different liquid crystalline classes. Indeed, the conventional rod-like and disc-

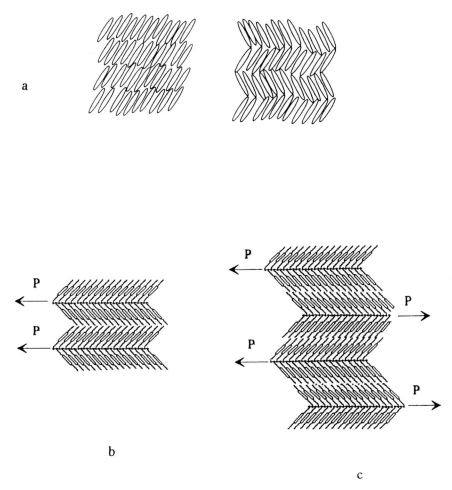

Fig. 17a–c. Sketches of the molecular arrangements for the smectic structure with alternating layer-to-layer tilt: **a** conventional and chevron smectic C layering in low molecular mass mesogens; **b** ferroelectric bilayer chevron structures for achiral side-chain polymers; **c** antiferroelectric bilayer chevron structures for achiral side-chain polymers. *Arrows* indicate the macroscopic polarization in the direction of the molecular tilt

like molecules are immiscible even in isotropic and nematic phases. The disc shaped molecules form the nematic phase with the long range order in the orientation of the normals to the discs planes. With decreasing temperature the more ordered columnar phases, in which stacked columns of molecules are arranged into a two-dimensional array, are formed. The two-dimensional lattices may be of hexagonal or rectangular symmetry and are characterized by long-range positional order of the columnar packing [94]. Within the columns, the positional order is liquid-like. Recent progress in synthesis of molecules combining the properties of rods and discs provided a new understanding on the relationship between columnar and lamellar ordering [18–20]. Actually there is no gap between these two classes of materials. One can proceed from the phases with one-dimensional periodicity (smectics) to phases with two-dimensional positional order (columnar phases) without phase separation by using appropriately designed mesogenic molecules. The aromatic part of the corresponding compounds is build up of at least five phenyl rings with suitable linkages. Two or three lengthy alkoxy chains are attached to ortho and para positions on each terminal phenyl ring (Fig. 18a). The term *phasmidic* is widely used for liquid crystals with a rod-like aromatic core and six aliphatic chains attached [18]. Other examples are the double biforked and double swallow-tailed mesogens [19, 20].

These compounds show a nematic and smectic C phase for relatively short chain lengths (n ≈ 7–10) and a hexagonal (columnar) phase for longer chains.

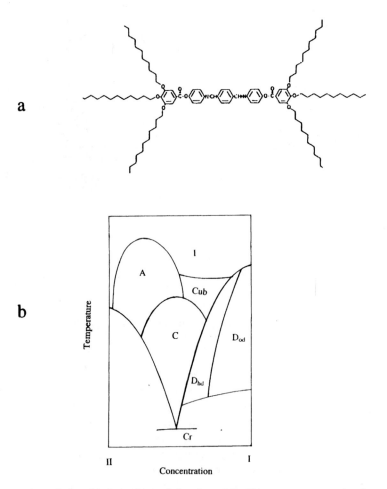

Fig. 18a,b. The polymorphic behaviour of the phasmidic-like mesogens: **a** molecular model with six terminal chains; **b** phase diagram for the binary mixture of double swallow-tailed compound (I) with conventional rod-like mesogen (II) (adapted from Letko et al. [123])

For mesogens with intermediate chain length, both the tilted lamellar and columnar phases have been observed. The columnar phases are made up from disc-like clusters of associated phasmidic molecules, which are stacked one upon another into columns [122]. The smectic A phase has not been found among phasmidics. However, it may be induced in certain binary mixtures of double swallow-tailed compounds with conventional rod-like mesogens [123]. The smectic A phase cannot transform directly into the columnar phase with two-dimensional periodicity, but only through an intermediate cubic phase (Fig. 18b). The cubic phase is an optically isotropic structure having a body centered lattice with a dimension of the order of the molecular length. The observed phase sequence is in accordance with the Landau-Brazovskii theory of weak crystallization for anisotropic systems [124], which clearly predicts a cubic (so-called "gyroid") phase in between the lamellar and columnar phases on the smectic – crystal phase diagrams. The phase behaviour of phasmidic-like mesogens resembles the rich polymorphism of lyotropic (micellar) systems where the amphiphilic molecules form lamellar or columnar (hexagonal) phases divided by the intermediate cubic phase upon variation of the water content [125]. The segregation between the aromatic cores and flexible alkylene chains in phasmidics is similar to the interaction between the amphiphilic molecules and the solvent in lyotropic systems. From this point of view, the molecular assembly in phasmidics is determined by the properties of the "interface" between the rigid molecular cores and the molten aliphatic chains [122].

The importance of the effects of segregation for the transition from the lamellar structure to columnar ordering is also readily seen for the case of the perfluorinated swallow-tailed compounds studied in [41a] (Fig. 11b). Upon cooling, a smectic A phase undergoes a transition to a columnar phase built up from rectangular clusters of overlapped perfluorinated molecules. The features of the two-dimensional packing of columns of such clusters are determined by the relative strength of the steric and polyphilic interactions leading to segregation of hydrocarbon and perfluorinated moieties.

7
Final Remarks

In this article a review has been given of particular types of structures in thermotropic liquid crystals formed by polar, sterically asymmetric and polyphilic molecules with different degrees of conformational freedom. In special cases, the formation of strongly defective layered structures and the effects of segregation of molecular fragments of different nature have been detected. Of course, the cases presented above do not exhaust the total possible variety of one and two-dimensional structures in liquid crystals. We have not touched here upon the chiral layered systems, which give a large number of defect (TGB) and antiferro- and ferrielectric phases. The structure of disc-like and mesogenic polymer phases, although very important, also remains beyond the scope of this article. Knowing the high creativity of organic chemists in the synthesis of new molecules of almost any shape and fragment sequences, we have no doubts that soon whole new types of liquid crystals will be synthesized, the properties of which are hard to imagine at the present time. We may expect the most interesting results for mesogenic molecules occupying the marginal states between the different types of liquid crystalline order such as the phasmids or biforked mesogens with semifluorinated chains. The strongly sterically asymmetric molecules (i.e. zig-zag, dumbell or banana shaped) possessing chemically different fragments are another object of interest. For these molecules, the effects of sublayer segregation, including the formation of polar types of ordering and particular tilted layered structures, are expected to show up clearly.

Acknowledgments. The author greatly appreciates collaboration and valuable discussions with L.M. Blinov, S. Diele, W.H. de Jeu, M.A. Osipov and F.G. Tournilhac. Special thanks are due to M. Bates for careful reading of the manuscript. This work was supported in part by the Netherlands Organization for Scientific Research (NWO).

8
References

1. For some recent reviews, see: (a) De Jeu WH (1992) In: Martelucci S, Chester AN (eds) Phase transitions in liquid crystals. Plenum Press, New York, chaps 1–3; (b) Osipov MA (1998) In: Goodby J, Demus D (eds) Handbook of liquid crystals, 2nd edn, Wiley-VCH, Berlin
2. Maier W, Saupe A (1959) Z Naturforschg 14a: 882; Maier W, Saupe A (1960) 15a: 287
3. Pohl L, Eidenschink R, Krause J, Weber G (1978) Phys Lett A 65: 169
4. Demus D, Demus H, Zaschke H (1983) Flüssige Kristalle in Tabellen, 2nd edn. VEB Verlag, Leipzig
5. (a) Frenkel D, Mulder BM (1985) Mol Phys 55: 1171; (b) Veerman JAC, Frenkel D (1990) Phys Rev A 41: 3237
6. (a) Poniwierski A, Sluckin TJ (1991) Phys Rev A 43: 6837; (b) Somoza AM, Tarazona P (1990) Phys Rev A 41: 965
7. Gelbart WM, Gelbart A (1977) Mol Phys 33: 1387
8. McMillan WL (1971) Phys Rev A 4: 1238; McMillan WL (1972) 6: 936
9. Kobayashi K (1970) J Phys Soc Japan 29: 101
10. See, for example, Wojtowicz PI (1979) In: Priestley PY, Wojtowicz PI, Sheng P (eds) Introduction to liquid crystals. Plenum Press, New York, chap 7 and references cited therein
11. Hardouin F, Levelut AM, Achard MF, Sigaud G (1983) J Chim Phys 80: 53 and references cited therein
12. Shashidhar R, Ratna BR (1989) Liq Cryst 5: 421 and references cited therein
13. Ostrovskii BI (1993) Liq Cryst 14: 131 and references cited therein
14. (a) Stroobants A, Lekkerkerker HNV, Frenkel D (1986) Phys Rev Lett 57: 1452; (b) Stroobants A, Lekkerkerker HNV, Frenkel D (1987) Phys Rev A 36: 2929; (c) Frenkel D (1988) J Phys Chem 92: 3280
15. (a) Mulder BM (1987) Phys Rev A 35: 3095; (b) Poniewierski A, Holyst R (1988) Phys Rev Lett 61: 2461; (c) Somoza AM, Tarazona P (1990) Phys Rev A 41: 965
16. Mederos L, Sullivan DE (1989) Phys Rev A 39: 854
17. (a) Luckhurst GR, Stephens RA, Phippen RW (1990) Liq Cryst 8: 451; (b) Luckhurst GR, Simmonds PSJ (1993) Mol Phys 80: 233
18. Malthete J, Levelut AM, Tinh NH (1985) J Phys Lett 46: 875
19. (a) Tinh NH, Destrade C, Levelut AM, Malthete J (1986) J Phys 47: 553; (b) Diele S, Ziebarth K, Pelzl G, Demus D, Weissflog W (1990) Liq Cryst 8: 211

20. Demus D (1989) Liq Cryst 5: 75
21. Attard GS, Garnett S, Hickman CC, Imrie CT, Taylor L (1990) Liq Cryst 7: 495
22. Shibaev VP, Lam L (eds) (1994) Liquid crystalline and mesomorphic polymers. Springer, Berlin Heidelberg New York
23. See, for example, Emsley JW (ed) (1985) Nuclear magnetic resonance in liquid crystals. Reidel, Dordrecht
24. (a) Poldy F, Dvolaitsky M, Taupin C (1975) J Phys Coll C1 36: 27; (b) Charvolin J, Deloche B (1979) In: Luckhurst GR, Gray GW (eds) The molecular physics of liquid crystals. Academic Press, London, chap 15
25. Luckhurst GR (1985) In: Chapoy LL (ed) Recent advances in liquid crystalline polymers. Elsevier, London, p105
26. Marcelja S (1974) J Chem Phys 60: 3599
27. (a) Samulski ET, Toriumi H (1983) J Chem Phys 79: 5194; (b) Janik B, Samulski ET, Toriumi H (1987) J Phys Chem 91: 1842
28. Emsley JW, Luckhurst GR, Stockley CP (1982) Proc R Soc London A381: 117
29. (a) Dowell F (1987) Phys Rev A 36: 5046; (b) Dowell F (1988) Phys Rev A 38: 382
30. (a) Cladis P (1975) Phys Rev Lett 35: 48; (b) Cladis PE, Bogardus RK, Aadsen D (1978) Phys Rev A 18: 2292
31. Tokita K, Fujimura K, Kondo S, Takeda M (1981) Mol Cryst Liq Cryst Lett 64: 171
32. (a) Indekeu JO, Berker AN (1986) Phys Rev A 33: 1158; (b) Indekeu JO, Berker AN (1986) Physica A 140: 368; (c) Netz RR, Berker AN (1992) Phys Rev Lett 68: 333
33. (a) Osman MA (1983) Z Naturf (a) 38: 693; (b) De Jeu WH (1983) Phil Trans R Soc A 309: 217
34. Petrov AG, Derzhanski A (1987) Mol Cryst Liq Cryst 151: 303
35. Meyer RB (1969) Phys Rev Lett 22: 918
36. Osipov MA (1984) Ferroelectrics 58: 305
37. Lobko TA, Ostrovskii BI, Pavluchenko AI, Sulianov SN (1993) Liq Cryst 15: 361
38. Diele S, Lose D, Kruth H, Pelzl G, Guittard F, Cambon A (1996) Liq Cryst 21: 603
39. Pelzl G, Latif I, Diele S Novak M, Demus D, Sackmann H (1986) Mol Cryst Liq Cryst 139: 333
40. Goring P, Pelzl G, Diele S, Delavier P, Siemensmayer K, Etzbach KH (1995) Liq Cryst 19: 629
41. (a) Lose D, Diele S, Pelzl G, Dietzmann E, Weissflog W, Liq Cryst (to be published); (b) Dietzmann E, Weissflog W, Markscheffel S, Jakli A, Lose D, Diele S (1996) Ferroelectrics 180: 59; (c) Pelzl G, Diele S, Lose D, Ostrovskii BI, Weissflog W (1997) Cryst Res Techn 32: 99
42. (a) Hardouin F, Achard MF, Jin J-I, Shin J-W, Yun Y-K (1994) J Phys 4: 627; (b) Hardouin F, Achard MF, Jin J-I, Yun Y-K (1995) J Phys 5: 927
43. See, for example, (a) Leadbetter AJ (1979) In: Luckhurst GR, Gray GW (eds) The molecular physics of liquid crystals. Academic Press, London, chap 13; (b) Pershan PS (1988) Structure of liquid crystalline phases. World Scientific, Singapore; (c) Ostrovskii BI (1989) Sov Sci Rev Sec A **12**(2): 86; (d) Shashidhar R (1992) In: Martelucci S, Chester AN (eds) Phase transitions in liquid crystals. Plenum Press, New York, chaps 15, 16
44. Blinov LM, Lobko TA, Ostrovskii BI, Sulianov SN, Tournilchac FG (1993) J Phys II 3: 1121
45. Ostrovskii BI, Tournilchac FG, Blinov LM, Haase W (1995) J Phys II 5: 979
46. Fontes E, Heiney PA, Haseltine JL, Smith AB (1986) J Phys 47: 1553
47. Lobko TA, Ostrovskii BI, Haase W (1992) J Phys II 2: 1195
48. Gramsbergen EF, de Jeu WH (1989) Liq Cryst 4: 449
49. Leadbetter AJ, Norris EK (1979) Molec Phys 38: 669. There are different contributions which give rise to a broadening σ of the molecular centre of mass distribution function f(z). The most important are the long-wave layer displacement thermal fluctuations and the individual motions of molecules having a random diffusive nature. The layer displacement amplitude depends on the magnitude of the elastic constants of smectics:

$\sigma^2 \sim (KB)^{-1/2}$, where B and K are the elastic constants for compression of the layers and for splay deformation, respectively

50. (a) Kirov N, Simova P (1984) Vibrational spectroscopy of liquid crystals. Bulgar Acad Sci, Sofia; (b) Blinov LM (1983) Electro-optical and magneto-optical properties of liquid crystals. Wiley, Chichester
51. Aver'yanov AM, Adomenas PV, Zhuikov VA, Zyryanov VYa (1986) Zh Eksp Teor Fiz 91: 552
52. (a) Kocot A, Kruk G, Wrzalic R, Vij JK (1992) Liq Cryst 12: 1005; (b) Kim KH, Ishikawa K, Takezoe H, Fukuda A (1995) Phys Rev E 51: 2166
53. (a) Blinov LM, Tournilhac F (1993) Mol Mats 3: 93, (b) Blinov LM, Tournilhac F (1993) 3: 169
54. (a) Prost J, Barois P (1983) J Chim Phys 80: 65; (b) Barois P, Prost J, Pommier J (1992) In: Lam L, Prost J (eds) Solitons in liquid crystals. Springer, Berlin Heidelberg New York, chap 6
55. See, for example, (a) Toulouse G (1977) Communs Phys 2: 115; (b) Janssen T, Janner A (1987) Adv Phys 36: 519; (c) Blinc R, Levanyuk AP (eds) (1986) Incommensurate phases in dielectrics. Elsevier; (d) Cladis PE (1988) Mol Cryst Liq Cryst 165: 85
56. (a) Longa L, de Jeu WH (1983) Phys Rev A 28: 2380; (b) Helfrich W (1987) J Phys 48: 291
57. (a) Tinh NH, Hardouin H, Destrade C (1982) J Phys 43: 1127; (b) Shashidhar R, Ratna BR, Surendranath V, Raja VN, Krishna Prasad S, Nagabhusan C (1985) J Phys Lett 46: 445
58. Prost J, Toner J (1987) Phys Rev A 36: 5008
59. (a) Cladis PE, Brand HR (1984) Phys Rev Lett 52: 2261; (b) Hardouin F, Achard MF, Tinh NH, Sigaud G (1986) Mol Cryst Liq Cryst Lett 3: 7
60. (a) Kumar S, Chen L, Surendranath V (1991) Phys Rev Lett 67: 322; (b) Patel P, Chen L, Kumar S (1993) Phys Rev E 47: 2643
61. Brownsey GJ, Leadbetter AJ (1980) Phys Rev Lett 44: 1608
62. Mang JT, Cull B, Shi Y, Patel P, Kumar S (1995) Phys Rev Lett 74: 4241
63. (a) Diele S, Pelzl G, Weissflog W, Demus D (1988) Liq Cryst 3: 1047; (b) Diele S, Manke S, Weissflog W, Demus D (1989) Liq Cryst 6: 301
64. Diele S, Roth K, Demus D (1986) Cryst Res Techn 21: 97
65. Weissflog W, Demus D, Diele S, Nitschke P, Wedler W (1989) Liq Cryst 5: 111 and references therein
66. Davidson P, Keller P, Levelut AM (1985) J Phys 46: 939
67. Endres BW, Ebert M, Wendorff JH, Reck B, Ringsdorf H (1990) Liq Cryst 7: 217
68. (a) Watanabe J, Nakata Y, Simizu K (1994) J Phys 4: 551; (b) Nakata Y, Watanabe J (1997) Pol Journ 29: 193
69. Longa L, de Jeu WH (1983) Phys Rev A 26: 1632
70. (a) Luckhurst GR, Timimi BA (1981) Mol Cryst Liq Cryst 64: 253; (b) Guillon D, Scoulios A (1984) J Phys 45: 607; (c) Madhusudana NV, Rajan J (1990) Liq Cryst 7: 31
71. Considering the pair of overlapping antiparallel molecules (dimer), we have not to think that each molecule interacts predominantly only with its neighbour. The actual holding time of the pairs is very short, the pairs are created and annihilated from point to point
72. Flory PJ (1969) Statistical mechanics of chain molecules. Wiley, London
73. Guillon D, Poeti G, Scoulios A, Fanelli E (1983) J Phys Lett 44: 491
74. See, for example, Bates FS, Fredrickson GH (1990) Ann Rev Phys Chem 41: 525
75. Diele S, Oelsner S, Kuschel F, Hisgen B, Ringsdorf H (1988) Mol Cryst Liq Cryst 155: 393
76. (a) Tournilhac F, Bosio L, Nicoid GF, Simon J (1988) Chem Phys Lett 145: 452; (b) Tournilhac F, Simon J (1991) Ferroelectrics 114: 283
77. Sigaud G, Nguyen HT, Achard MF, Twieg RJ (1990) Phys Rev Lett 65: 2796 and references cited therein. We note that alkyl and fluoroalkyl groups are nonpolar and interact primarily via dispersion (Van der Waals) forces: V_{FH}-$\alpha_F\alpha_H$. Thus the

incompatibility of these fragments are due to differences in their polarizabilities α_F and $\alpha_{||}$. However, for dense fluids the effects of packing entropy may also be important

78. Tournilhac F, Blinov LM, Simon J, Yablonsky SV (1992) Nature 359: 621
79. Prost J, Bruinsma R, Tournilhac F (1994) J. Phys. II France 4: 169
80. (a) Petschek RG, Wiefling KM (1987) Phys Rev Lett 59: 343; (b) Perchak DR, Petschek RG (1991) Phys Rev A 43: 6756
81. See, for example, (a) Titov VV, Zverkova TI, Kovshov EI, Fialkov YuN, Shelazhenko SV, Yagupolski LM (1978) Mol Cryst Liq Cryst 47: 1; (b) Ivashcenko AV, Kovshov EI, Lazareva E, Prudnikova K, Titov VV, Zverkova TI, Barnik MI, Yagupolski LM (1981) Mol Cryst Liq Cryst 67: 235; (c) Koden M, Nakagawa K, Ishii, Y, Funada F, Matsumura M, Awane K (1989) Mol Cryst Liq Cryst Lett 6: 185; (d) Nguyen HT, Sigaud G, Achard MF, Hardouin F, Twieg RJ, Betterton K (1991) Liq Cryst 10: 389
82. (a) Janulis EP, Novack JC Papapolymerou GA, TristanKendra M, Huffman WA (1988) Ferroelectrics 85: 375; (b) Doi T, Sakurai Y, Tamatani A, Takenaka S, Kusabayashi S, Nishihata Y, Terauchi H (1991) J Mater Chem 1: 169
83. (a) Diele S (1993) Ber Bunsenges Phys Chem 97: 1326; (b) Kaganer VM, Diele S, Ostrovskii BI, Haase W (1997) Mol Mats 9: 59
84. (a) Stoebe T, Mach P, Huang CC (1994) Phys Rev Lett 73: 1384; (b) Johnson PM, Mach P, Wedell ED, Lintgen F, Neubert M, Huang CC (1997) Phys Rev E 55: 4386
85. Pang J, Clark NA (1994) Phys Rev Lett 73: 2332
86. Shindler JD, Mol EAL, Shalaginov A, de Jeu WH (1995) Phys Rev Lett 74: 722
87. (a) Rieker TP, Janulis EP (1994) Liq Cryst 17: 681; (b) Rieker TP, Janulis EP (1995) Phys Rev E 52: 2688
88. (a) Sereda SV, Antipin Mu, Timofeeva TV, Struchkov YuT, Shelazhenko SV (1987) Kristallografiya 32: 352; (b) Kromm P, Cotrait M, Roullon JC, Barois P, Nguyen HT (1997) Liq Cryst 21: 121
89. Pavluchenko AI, Smirnova NI, Petrov VF, Fialkov YuN, Shelazhenko SV, Yagupolski LM (1991) Mol Cryst Liq Cryst 209: 225
90. Stoebe T, Mach P, Grantz S, Huang CC (1996) Phys Rev E 53: 1662. In this work the surface tension of the free standing films of some perfluorinated compounds has been measured. The intermediate value of the surface tension ($\approx$14 dyn/cm) between that of pure close-packed CF_3 and CH_3 groups indicates that the surface of the film consist of nearly equal proportions of these fragments
91. (a) Rabolt JF, Russel TP, Twieg RJ (1984) Macromolecules 17: 2786; (b) Viney C, Twieg RJ, Russel TP, Depero LE (1989) Liq Cryst 5: 1783; (c) Hopken J, Moler M (1992) Macromolecules 25: 2482
92. See, for example, Kissa E (1994) Fluorinated surfactants. M Dekker, New York
93. (a) Mahler W, Guillon D, Scoulios A (1985) Mol Cryst Liq Cryst Lett 2: 111; (b) Shin S, Collazo N, Rice SA (1992) J Chem Phys 96: 1352
94. (a) Levelut AM (1983) J Chim Phys 80: 149; (b) Fontes E, Heiney PA, Ohba M, Haseltine JN, Smith AB (1988) Phys Rev A 37: 1329
95. Tournilhac F, Kumar S, private communication
96. Goodby JW, Waugh MA, Stein SM, Chin E, Pindak R, Patel JS (1989) J Am Chem Soc 111: 8119
97. Ibn Elhaj M, Coles HJ, Guillon D, Scoulios A (1993) J Phys II 3: 1807
98. Bartolino R, Doucet J, Durand G (1978) Ann Phys (Paris) 3: 389
99. Keller EN, Nachaliel E, Davidov D, Boffel Ch (1986) Phys Rev A 34: 4363
100. (a) McMillan WL (1973) Phys Rev A 8: 1921; (b) Van der Meer BW, Vertogen G (1979) J Phys Coll C3 40: 222; (c) Cabib D, Benguigui L (1977) J Phys 38: 419; (d) Goossens WJA (1985) J Phys 46: 1411
101. (a) Wulf A (1975) Phys Rev A 11: 365; (b) Sirota EB (1988) J Phys 49: 1443
102. The experimental data, especially for ferroelectric chiral systems, show that in tilted phases the molecules are still rotating around their long axis, though there are some steric hindrance; see, for example, Fukuda A, Takanishi Y, Isozaki T, Ishikava K, Takezoe H (1994) J Mater Chem 4: 997 and references cited therein

103. de Jeu WH (1977) J Phys France 38: 1265 and references cited therein
104. Barbero G, Durand G (1990) Mol Cryst Liq Cryst 179: 157
105. (a) Poniewierski A, Sluckin TJ (1991) Mol Phys 73: 199; (b) Velasko E, Mederos L, Sluckin TJ (1996) Liq Cryst 20: 399
106. (a) Glaser MA, Malzbender R, Clark NA, Walba DM (1994) J Phys Condens Matt A 6: 261; (b) Glaser MA, Malzbender R, Clark NA, Walba DM (1995) Mol Sim 14: 343
107. (a) Weissflog W, Pelzl G, Wiegeleben A, Demus D (1980) Mol Cryst Liq Cryst Lett 56: 295; (b) Tinh NH, Hardouin F, Deastrade C, Levelut AM (1982) J Phys Lett 43: 33
108. (a) Swager TM, Serrette AG, Knawby DM, Zheng H (1994) 15 ILCC, Budapest, Abstracts, 2: 771; (b) Xu B, Swager TM (1995) J Am Chem Soc 117: 5011
109. Niori T, Sekine T, Watanabe J, Furukawa T, Takezoe H (1996) J Mater Chem 6: 1231
110. Goodby JW, Clark NA, Lagerwall S, Osipov MA, Pikin SA (eds) (1991) Ferroelectric liquid crystals. Gordon and Breach, Philadelphia
111. Meyer RB, Liebert L, Strzelecki L, Keller P (1975) J Phys Lett 36: 69
112. Hiji N, Chandani ADL, Nishiyama S, Ouchi Y, Takezoe H, Fukuda A (1988) Ferroelectrics 85: 99
113. In bilayer A_2 phase the dipole arrangement is antiferroelectric-like. However there are no easy way to reverse the direction of dipoles by external field
114. (a) Lin Lei (1987) Mol Cryst Liq Cryst 146: 41; (b) Pallfy-Muhoray P, Lee MA, Petschek RG (1988) Phys Rev Lett 60: 2303; (c) Biscarini F, Zannoni C, Chiccoli C, Pasini P (1991) Mol Phys 73: 439; (d) Weis JJ, Levesque D, Zarragoicoechea GJ (1992) Phys Rev Lett 69: 913; (e) Ayton G, Wei DQ, Patey GN Phys Rev E 55: 447
115. Levesque D, Weis JJ, Zarragoicoechea GJ (1993) Phys Rev E 47: 496
116. Link DR, Natale G, Shao R, Maclennan JE, Clark NA, Korblova E, Walba DM (1997) Science 278: 1924
117. (a) Brand HR, Cladis P, Pleiner H (1992) Macromolecules 25: 7223; (b) Cladis P, Brand HR (1993) Liq Cryst 14: 1327
118. (a) Beresnev LA, Blinov LM, Baikalov VA, Pozhidayev EP, Purvanetskas GV, Pavluchenko AI (1982) Mol Cryst Liq Cryst 89: 327; (b) Levelut AM, Germain C, Keller P, Liebert L, Billard J (1983), J Phys 44: 623; (c) Watanabe J, Hayashi M (1989) Macromolecules 22: 4083; (d) Galerne Y, Liebert L (1990) Phys Rev Lett 64: 906
119. (a) Nishiyama I, Goodby JW (1992) J Mater Chem 2: 1015; (b) Heppke G, Kleinberg P, Lotzsch D (1993) Liq Cryst 14: 67
120. (a) Soto Bustamante EA, Yablonskii SV, Ostrovskii BI, Beresnev LA, Blinov LM, Haase W (1996) Chem Phys Lett 260: 447; (b) Soto Bustamante EA, Yablonskii SV, Ostrovskii BI, Beresnev LA, L Blinov LM, Haase W (1996) Liq Cryst 21: 829; (c) Ostrovskii BI, Soto Bustamante EA, Sulyanov SN, Galyametdinov YuG, Haase W (1996) Mol Mats 6: 171
121. Chandrasekhar S, Shadashiva BK, Suresh KA (1977) Pramana 9: 471
122. (a) Guillon D, Skoulios A, Malthete J (1987) Europhys Lett 3: 67; (b) Hendrikx Y, Levelut AM (1988) Mol Cryst Liq Cryst 165: 233
123. (a) Letko I, Diele S, Pelzl G, Weissflog W (1995) Liq Cryst 19: 643; (b) Letko I, Diele S, Pelzl G, Weissflog W (1995) Mol Cryst Liq Cryst 260: 171
124. (a) Kats EI, Lebedev VV, Muratov AR (1989) Physica A 160: 98; (b) Podneks VE, Hamley IW (1996) Pis'ma v ZhETF 64: 564
125. Charvolin J (1989) In: Riste T, Sherrington D (eds) Phase transitions in soft condensed matter. Plenum Press, New York, p95

Author Index Volumes 1–94

Baldwin AH, see Butler A (1997) 89: 109–132

Balsenc LR (1980) Sulfur Interaction with Surfaces and Interfaces Studied by Auger Electron Spectrometry. 39: 83–114

Banci L, Bencini A, Benelli C, Gatteschi D, Zanchini C (1982) Spectral-Structural Correlations in High-Spin Cobalt(II) Complexes. 52: 37–86

Banci L, Bertini I, Luchinat C (1990) The ^{1}H NMR Parameters of Magnetically Coupled Dimers-The Fe_2S_2 Proteins as an Example. 72: 113–136

Baran EJ, see Müller A (1976) 26: 81–139

Bartolotti LJ (1987) Absolute Electronegatives as Determined from Kohn-Sham Theory. 66: 27–40

Bates MA, Luckhurst GR (1999) Computer Simulation of Liquid Crystal Phases Formed by Gay-Berne Mesogens. 94: 65–137

Bau RG, see Teller R (1981) 44: 1–82

Baughan EC (1973) Structural Radii, Electron-cloud Radii, Ionic Radii and Solvation. 15: 53–71

Bayer E, Schretzmann P (1967) Reversible Oxygenierung von Metallkomplexen. 2: 181–250

Bearden AJ, Dunham WR (1970) Iron Electronic Configuration in Proteins: Studies by Mössbauer Spectroscopy. 8: 1–52

Bencini A, see Banci L (1982) 52: 37–86

Benedict U, see Manes L (1985) 59/60: 75–125

Benelli C, see Banci L (1982) 52: 37–86

Benfield RE, see Thiel RC (1993) 81: 1–40

Bergmann D, Hinze J (1987) Electronegativity and Charge Distribution. 66: 145–190

Berners-Price SJ, Sadler PJ (1988) Phosphines and Metal Phosphine Complexes: Relationship of Chemistry to Anticancer and Other Biological Activity. 70: 27–102

Bertini I, see Banci L (1990) 72: 113–136

Bertini I, Ciurli S, Luchinat C (1995) The Electronic Structure of FeS Centers in Proteins and Models. A Contribution to the Understanding of Their Electron Transfer Properties. 83:1–54

Bertini I, Luchinat C, Scozzafava A (1982) Carbonic Anhydrase: An Insight into the Zinc Binding Site and into the Active Cavity Through Metal Substitution. 48: 45–91

Bertrand P (1991) Application of Electron Transfer Theories to Biological Systems. 75: 1–48

Bill E, see Trautwein AX (1991) 78: 1–96

Bino A, see Ardon M (1987) 65: 1–28

Blanchard M, see Linarès C (1977) 33: 179–207

Blasse G, see Powell RC (1980) 42: 43–96

Blasse G (1991) Optical Electron Transfer Between Metal Ions and its Consequences. 76: 153–188

Blasse G (1976) The Influence of Charge-Transfer and Rydberg States on the Luminescence Properties of Lanthanides and Actinides. 26: 43–79

Blasse G (1980) The Luminescence of Closed-Shell Transition Metal-Complexes. New Developments. 42: 1–41

Blauer G (1974) Optical Activity of Conjugated Proteins. 18: 69–129

Bleijenberg KC (1980) Luminescence Properties of Uranate Centres in Solids. 42: 97–128

Boca R, Breza M, Pelikán P (1989) Vibronic Interactions in the Stereochemistry of Metal Complexes 71: 57–97

Boeyens JCA (1985) Molecular Mechanics and the Structure Hypothesis. 63: 65–101

Böhm MC, see Sen KD (1987) 66: 99–123

Bohra R, see Jain VK (1982) 52: 147–196

Bollinger DM, see Orchin M (1975) 23: 167–193

Bominaar EL, see Trautwein AX (1991) 78: 1–96

Bonnelle C (1976) Band and Localized States in Metallic Thorium, Uranium and Plutonium, and in Some Compounds, Studied by X-ray Spectroscopy. 31: 23–48

Bose SN, see Nag K (1985) 63: 153–197

Bowler BE, see Therien MJ (1991) 75: 109–130

Ibers JA, Pace LJ, Martinsen J, Hoffmann BM (1982) Stacked Metal Complexes: Structures and Properties. 50: 1-55

Ingraham LL, see Maggiora GM (1967) 2: 126-159

Iqbal Z (1972) Intra- and Inter-Molecular Bonding and Structure of Inorganic Pseudohalides with Triatomic Groupings. 10: 25-55

Izatt RM, Eatough DJ, Christensen JJ (1973) Thermodynamics of Cation-Macrocyclic Compound Interaction. 16: 161-189

Jain VK, Bohra R, Mehrotra RC (1982) Structure and Bonding in Organic Derivatives of Antimony(V). 52: 147-196

Jerome-Lerutte S (1972) Vibrational Spectra and Structural Properties of Complex Tetracyanides of Platinum, Palladium and Nickel. 10: 153-166

Johnston RL (1997) Mathematical Cluster Chemistry. 87: 1-34

Johnston R, see Mingos DMP (1987) 68: 29-87

Jolivet JP, see Henry M (1991) 77: 23-47

Jørgensen CK, see Müller A (1973) 14: 153-206

Jørgensen CK, see Reisfeld R (1982) 49: 1-36

Jørgensen CK, see Reisfeld R (1988) 69: 63-96

Jørgensen CK, see Reisfeld R (1991) 77: 207-256

Jørgensen CK, Frenking G (1990) Historical, Spectroscopic and Chemical Comparison of Noble Gases. 73: 1-16

Jørgensen CK, Kauffmann GB (1990) Crookes and Marignac - A Centennial of an Intuitive and Pragmatic Appraisal of "Chemical Elements" and the Present Astrophysical Status of Nucleosynthesis and "Dark Matter". 73: 227-254

Jørgensen CK, Reisfeld R (1982) Uranyl Photophysics. 50: 121-171

Jørgensen CK (1976) Deep-Lying Valence Orbitals and Problems of Degeneracy and Intensitites in Photo-Electron Spectra. 30: 141-192

Jørgensen CK (1966) Electric Polarizability, Innocent Ligands and Spectroscopic Oxidation States. 1: 234-248

Jørgensen CK (1990) Heavy Elements Synthesized in Supernovae and Detected in Peculiar A-type Stars. 73: 199-226

Jørgensen CK (1996) Luminescence of Cerium (III) Inter-Shell Transitions and Scintillator Action. 85: 195-214

Jørgensen CK (1976) Narrow Band Thermoluminescence (Candoluminescence) of Rare Earths in Auer Mantles. 25: 1-20

Jørgensen CK (1975) Partly Filled Shells Constituting Anti-bonding Orbitals with Higher Ionization Energy than Their Bonding Counterparts. 22: 49-81

Jørgensen CK (1975) Photo-Electron Spectra of Non-Metallic Solids and Consequences for Quantum Chemistry. 24: 1-58

Jørgensen CK (1978) Predictable Quarkonium Chemistry. 34: 19-38

Jørgensen CK (1966) Recent Progress in Ligand Field Theory. 1: 3-31

Jørgensen CK (1967) Relationship Between Softness, Covalent Bonding, Ionicity and Electric Polarizability. 3: 106-115

Jørgensen CK (1981) The Conditions for Total Symmetry Stabilizing Molecules, Atoms, Nuclei and Hadrons. 43: 1-36

Jørgensen CK (1973) The Inner Mechanism of Rare Earths Elucidated by Photo-Electron Spectra. 13: 199-253

Jørgensen CK (1969) Valence-Shell Expansion Studied by Ultra-violet Spectroscopy. 6: 94-115

Justin KR, see Chandrasekhar V (1993) 81: 41-114

Kadish KM, see Guilard R (1987) 64: 205-268

Kahn O (1987) Magnetism of the Heteropolymetallic Systems. 68: 89-167

Kalyanasundaram K, see Kiwi J (1982) 49: 37-125

Kauffmann GB, see Jørgensen CK (1990) 73: 227-254

Printed by Publishers' Graphics LLC
CAMZ131121.15.22.750